High Performance Computing in the Geosciences

NATO ASI Series

Advanced Science Institutes Series

A Series presenting the results of activities sponsored by the NATO Science Committee, which aims at the dissemination of advanced scientific and technological knowledge, with a view to strengthening links between scientific communities.

The Series is published by an international board of publishers in conjunction with the NATO Scientific Affairs Division

A Life Sciences **B Physics**	Plenum Publishing Corporation London and New York
C Mathematical and Physical Sciences **D Behavioural and Social Sciences** **E Applied Sciences**	Kluwer Academic Publishers Dordrecht, Boston and London
F Computer and Systems Sciences **G Ecological Sciences** **H Cell Biology** **I Global Environmental Change**	Springer-Verlag Berlin, Heidelberg, New York, London, Paris and Tokyo

PARTNERSHIP SUB-SERIES

1. Disarmament Technologies	Kluwer Academic Publishers
2. Environment	Springer-Verlag / Kluwer Academic Publishers
3. High Technology	Kluwer Academic Publishers
4. Science and Technology Policy	Kluwer Academic Publishers
5. Computer Networking	Kluwer Academic Publishers

The Partnership Sub-Series incorporates activities undertaken in collaboration with NATO's Cooperation Partners, the countries of the CIS and Central and Eastern Europe, in Priority Areas of concern to those countries.

NATO-PCO-DATA BASE

The electronic index to the NATO ASI Series provides full bibliographical references (with keywords and/or abstracts) to more than 50000 contributions from international scientists published in all sections of the NATO ASI Series.
Access to the NATO-PCO-DATA BASE is possible in two ways:

– via online FILE 128 (NATO-PCO-DATA BASE) hosted by ESRIN, Via Galileo Galilei, I-00044 Frascati, Italy.

– via CD-ROM "NATO-PCO-DATA BASE" with user-friendly retrieval software in English, French and German (© WTV GmbH and DATAWARE Technologies Inc. 1989).

The CD-ROM can be ordered through any member of the Board of Publishers or through NATO-PCO, Overijse, Belgium.

Series C: Mathematical and Physical Sciences – Vol. 462

High Performance Computing in the Geosciences

edited by

François-Xavier Le Dimet
Laboratoire de Modélisation et Calcul,
Université Joseph Fourier,
Grenoble, France

Springer-Science+Business Media, B.V.

Proceedings of the NATO Advanced Research Workshop on
High Performance Computing in the Geosciences
Centre de Physique, Les Houches, France
21–25 June 1993

A C.I.P. Catalogue record for this book is available from the Library of Congress.

ISBN 978-94-010-4019-8 ISBN 978-94-011-0033-5 (eBook)
DOI 10.1007/978-94-011-0033-5

Printed on acid-free paper

To A. T.

"La queue de l'âne était une ficelle tressée et une légende expliquait comment cette queue pouvait servir de baromètre :

If tail is dry	*Fine*
If tail is wet	*Rain*
If tail moves	*Windy*
If tail cannot be seen	*Fog*
If tail is frozen	*Cold*
If tail falls out	*Earthquake*

Le pub était bondé..."

Georges Perec
La Vie mode d'emploi

Hachette, 1978

Table of Contents

Preface

This book is an outcome of the 1994 Workshop on "High Performance Computing in the Geosciences" held at Centre de Physique in Les Houches, France.

National and international programs for studying planet Earth over extended spectra of spatial and temporal scale require an ever-increasing use of numerical modeling for the understanding and the prediction of various planetary phenomena at every resolvable scale.

An important development of numerical modeling in the geosciences may be justified by the following :

1. An intrinsic characteristic of the planetary sciences is the non-reproducibility of in situ experiments. In other area of science where phenomena are governed by non-linear partial differential equations the validation of models and their tuning may be obtained by means of physical experiments. Numerical modeling should allow to partially alleviate the non-reproducibility handicap however at a high computational cost.

2. Presently a multitude of spatially observing platforms such as EOS are being deployed and will result an increase by a factor of 100 of the volume of observational data. Without an underlying numerical model permitting a coherent reconstitution of the different fields this additional mass of data might be of little use. High performance numerical methods must be employed for the optimal use of this considerable amount of additional information.

3. Our understanding of planet earth is becoming increasingly global both at the level of its components (atmosphere, ocean, biosphere and solid earth) as well as at the level of space and time scales. The implementation of coupled models (e.g. ocean-atmosphere) becomes mandatory for the comprehension of an increasing number of phenomena. The models that one would have to consider will be of huge dimensionality which at the present state of the art of computational tools at our disposal will result in a prohibitively high computational cost.

4. The mathematical tools, such as dynamical systems or optimal control theory, used for explaining, describing and predicting the behavior of planetary components or their interactions allow a finer approach to numerical modeling. The use of these tools proves to be indispensable for the understanding of a number of phenomena but their implementation requires a considerable additional computational effort.

5. Presently, one observes the emergence of a strong social pressure requiring the availability of more accurate forecasts, both for the classical numerical weather prediction as well as for the forecast of catastrophic events and for the impact of anthropogenic activities on the environment.

The development required by the numerical modeling of physical-chemical phenomena in the ocean and the atmosphere also requires considerable computational means.

The major obstacle encountered in the majority of the aforementioned problems is due to the limitation of computational capability. Even if one were to extrapolate the resources available on present day computers to a relatively near future one finds that we are still far away from attaining the computational capability required for solving these problems.

One possible solution for alleviating this problem is to recourse to massively parallel processing (MPP).

The changes induced by porting these models to MPP's go further than purely technological aspects and require a rather different approach to numerical modeling. Porting a complex numerical model to MPP's constitute a very expensive investment both in manpower and equipment. Initial methodological choices may become rapidly irreversible and the model risks to attain obsolescence even before its operational implementation.

The aims of this workshop were:

- Survey the state of the art of programs presently being developed and which require High Performance Computing for their implementation.
- Provide a guide for decision making in what concerns computing directions in future numerical models.

- Provide an ensemble view of future developments of MPP's and their implications for numerical modeling in the geosciences.

This meeting has been made possible with the help of :

- Université Joseph Fourier in Grenoble
- Institut Informatique et Mathématiques Appliquées de Grenoble
- Direction de la Recherche et des Etudes Techniques (DGA)
- NATO Scientific Division
- Ministère des Affaires Etrangères.

My personal thanks go to Professor I.-M. NAVON from the Supercomputer Computation Research Institute at Florida State University for his constant and friendly support. Yannick Trémolet has been a very efficient adjoint that I would like to thanks.

Grenoble, January 1995

François-Xavier LE DIMET
Laboratoire de Modélisation et Calcul
Université Joseph Fourier, Grenoble, France
Projet IDOPT CNRS-INRIA

List of Contributors

Alekseev A.S.
Computing Center of the Siberian Division of the Russian Academy of Sciences
Lavrentiev Prospects 6
Novosibirrsk 630090
Russia

Atlas Susan R.
Center for Nonlinear Studies
MS B 258
Los Alamos National Laboratory
Los Alamos NM 87545
USA

Baillie C.F.
NOAA Forecast Systems Laboratory
Boulder CO 80303
USA

Bernier Christine
Département de Mathématiques
Université de Nancy 1
BP 239
F-54506 Vandoeuvre-lès-Nancy Cedex
France

Bischof Christian H.
Mathematics and Computer Sciences Division
Argonne National Laboratory
9700 S. Cass. Ave.
Argonne IL 60439-4843
USA

Cook Jeremy
Parallab, Deparment of Informatics
University of Bergen
N-5020 Bergen
Norway

Dash Sushil Kumar
Indian Institute of Technology
Centre for Atmospheric Sciences
Hauz Khas
New Delhi-110016
India

Gravel Sylvie
Recherche en Prévision Numérique
Environnement Canada
2121 Route Transcanadienne
Dorval Québec H9P 1J3
Canada

Guyon Marc
CETIIS
24 Bd Paul Vaillant-Couturier
F-94200 Ivry-sur-Seine
France

Navon I.M.
Department of Mathematics and Supercomputer Computations Research Institute
Florida State University
Tallahassee FL 32306-4052
USA

O'Keefe Matthew
Department of Electrical Engineering
University of Minnesota
200 Union St. S.E.
Minneapolis MN 55455
USA

Rodriguez Bernardo
Advanced Computing Group
NOAA Forecast Systems Laboratory
R/E/FS5
325 Broadway
Boulder CO 80303
USA

Rostaing-Schmidt Nicole
Université de Nice-Sophia Antipolis
Parc Valrose
BP 71
F-06108 Nice Cedex 2
France

Skalin Roar
SINTEF Industrial Mathematics
N-7034 Trondheim
Norway

Terray Laurent
Cerfacs
42 Avenue G. Coriolis
F-31057 Toulouse Cedex
France

Tolstykh Mikhail A.
Institute of Numerical Mathematics
Russian Academy of Sciences
Leninskii prospects 32A
Moscow 117334
Russia

Wollenweber Fritz G.
European Center for Medium Range Weather Forecasts
Shinfield Park
Reading Berkshire RG2 9AX
UK

List of Participants

Ababou Rachid
DRN/DMT/SEMT/TTMF
CEA SACLAY
F-91191 Gif-sur-Yvette
France

Alekseev Anatoly
Mathematical Problems of Geophysics
Russian Academy of Sciences
Lavrentiev Prospects 6
Novossibirsk 630090
Russia

Alekseev Vladimir
Institute of Numerical Mathematics
Russian Academy of Sciences
Leninskii Prospects 32A
Moscow 117334
Russia

Ashworth Mike
Bidston Observatory
NERC Computer Service
Birkenhead L43 7RA
UK

Atlas Susan R.
Thinking Machine Corporation
Los Alamos National Laboratory
Los Alamos NM 87545
USA

Bagneres Anne
LMC
Université Joseph Fourier
BP 53X
F-38041 Grenoble Cedex
France

Baillie Clive
Physics Department
University of Colorado
Boulder CO 80309
USA

Berges Jean-Claude
TE/IS/MS/MN
CNES
18 Avenue Edouard Belin
F-31055 Toulouse Cedex
France

Bernier Christine
Départment de Mathématiques
Université de Nancy 1
BP 239
F-54506 Vandoeuvre-lès-Nancy Cedex
France

Bischof Christian H.
Mathematics and Computer Sciences Division
Argonne National Laboratory
9700 S. Cass. Ave.
Argonne IL 60439
USA

Bjorge Dag
The Norwegian Meteorological Institute
PO Box 43
N-0313 Oslo
Norway

Blayo Eric
LEGI, IMG
BP 53X
F-38041 Grenoble Cedex
France

Budich Reinhard
Abteilung Theoretische Klimamodellierung
Max-Planck Institute für Meteorologie
Bundesstr. 55
D-2000 Hamburg 13
Germany

Bureau Stephane
TELMAT
6 rue de l'Industrie
BP 12
F-68360 Soultz
France

Cekirge H.M.
Geophysical Fluid Dynamics
Florida State University
Tallahassee FL 32306-4052
USA

Clement Philippe
Laboratoire de Modélisation
CEA-CENG
BP 85X
F-38041 Grenoble
France

Cohen Claude
Cray Research
7 rue de Tilsitt
F-75017 Paris
France

Cubasch Ulrich
Deutches Klimarechenzentrum
Bundesstr. 55
D-2000 Hamburg 13
Germany

Dash Sushil Kumar
Indian Institute of Technology
Center for Atmospheric Sciences
Hauz Khas
New Delhi 110016
India

Drogenmeier Kelvin
School of Meteorology
University of Oklahoma
Energy Center
Norman OK 73019
USA

Foster Ian
Mathematics and Computer Sciences Division
Argonne National Laboratory
9700 S. Cass. Ave.
Argonne IL 60439
USA

Francis Rhys
High Performance Computing
CSIRO
723 Swanston St.
Carlton 3035
Australia

Galligo André
Laboratoire de Mathématiques
Université de Nice-Sophia Antipolis
Parc Valrose
F-06108 Nice Cedex 2
France

Gartel Ute
GMD GmbH
PO Box 1316
Schloss Birlinghoven
D-5205 St. Augustin 1
Germany

Ghemires Mohamed
LMC
Université Joseph Fourier
BP 53X
F-38041 Grenoble Cedex
France

Gravel Sylvie
Atmospheric Environment Service
Environenment Canada
2121 Route Transcanadienne
Dorval Québec H9P 1J3
Canada

Griewank Andreas
Mathematics and Computer Sciences Division
Argonne National Laboratory
S. Cass. Ave.
Argonne IL 60439
USA

Gulzow Volker
Deutches Klimarechenzentrum
Bundesstr. 55
D-2000 Hamburg 13
Germany

Guyon Marc
ONERA / DMI
BP 72
29 Ave. de la Division Leclerc
F-92320 Chatillon
France

Hassold Eric
Laboratoire de Mathématiques
Université de Nice-Sophia Antipolis
Parc Valrose
F-06108 Nice Cedex 2
France

Henderson James
Computer Centre
University of London
20 Guilford St.
London WC1N 1DZ
UK

Jami Adrien
Direction de Etudes et Recherches
EDF
1 Ave. Général De Gaulle
F-92141 Clamart Cedex
France

Jude Jonathan
TELMAT
6 rue de l'Industrie
BP 12
F-68360 Soultz
France

Kauranne Tuomo
Department of Mathematics
University of Joensuu
PO Box 111
SF 80101 Joensuu
Finland

Kowalski A. Daniel
Universities Space Research Assoc.
NASA Goddard Space Flight Center
Greenbelt MD 20771
USA

Latour Jean
Cray Research France
Meteo-France ISCEM/TTI/DIV
42 Avenue G. Coriolis
F31057 Toulouse Cedex
France

Le Dimet François-Xavier
LMC
Université Joseph Fourier
BP 53X
F-38941 Grenoble Cedex
France

Lie Ivar
Division for Electronics
Norwegian Defense Research
Establishment
PO Box 25
N-2007 Kjeller
Norway

Mascart Patrick
Laboratoire d'Aéréologie
Université Paul Sabatier
118 route de Narbonne
F-31062 Toulouse
France

McRae Greg
Department of Chemical Engineering
Massachusetts Institute of Technology
25 Ames St.
Cambridge MA 02139
USA

Navon I. Mickael
Supercomputer Computations Research
Institute
Florida State University
Tallahassee FL 32306-4052
USA

O'Keefe Matthew
Department of Electrical Engineering
University of Minnesota
200 Union St. S.E.
Minneapolis MN 55455
USA

Rodriguez Bernardo
NOAA/FLS
Mailstop R/E/F05
325 Broadway
Boulder CO 80303
USA

Rosset Robert
Laboratoire d'Aérologie
Université Paul Sabatier
118 Route de Narbonne
F-31062 Toulouse Cedex
France

Rostaing-Schmidt Nicole
Laboratoire de Mathématiques
Université de Nice-Sophia Antipolis
Parc Valrose
F-06108 Nice Cedex 2
France

Salmond Deborah
Cray Research
Oldbury
Bracknell Berkshire
UK

Simmons Adrian
ECMWF
Shinfield Park
Reading Berkshire RG2 9AX
UK

Skalin Roar
SINTEF Industrial Mathematics
N-7034 Trondheim
Norway

Suhre Karsten
Laboratoire d'Aérologie
Université Paul Sabatier
118 Route de Narbonne
F-31062 Toulouse Cedex
France

Swarztrauber Paul N.
NCAR
PO Box 3000
1850 Table Mesa Drive
Boulder CO 80307
USA

Terray Laurent
Climate Modeling and Global Change Team
Cerfacs
42 Avenue G. Coriolis
F-31057 Toulouse Cedex
France

Thomas Steve
Recherche en Prévision Numérique
Service de l'environnement atmosphérique
1314 Crawford Bridge
Verdun Québec H4H 2N5
Canada

Thomas Bernadette
Direction de Etudes et Recherches
EDF
1 Ave. Général De Gaulle
F-92141 Clamart Cedex
France

Tolstykh Mikhail A.
Institute of Numerical Mathematics
Russian Academy of Sciences
Leninskii Prospects 32A
Moscow 117334
Russia

Tremolet Yannick
LMC, IMAG
46 Ave. Félix Viallet
F-38031 Grenoble Cedex
France

Verron Jacques
LEGI, IMG
BP 53X
F-38041 Grenoble Cedex
France

Wollenweber Fritz G.
ECMWF
Shinfield Park
Reading Berkshire RG2 9AX
UK

PERFORMANCE AND PORTABILITY IN PARALLEL COMPUTING: A WEATHER FORECAST VIEW

BERNARDO RODRIGUEZ
LESLIE HART
TOM HENDERSON
Advanced Computing Group
Forecast Systems Laboratory
National Oceanic and Atmospheric Administration
R/E/FS5
325 Broadway
Boulder, CO. 80303

ABSTRACT. We have developed a high level library, the Nearest Neighbor Tool (NNT), to facilitate the coding of finite difference approximation weather prediction models on parallel computers. NNT provides portability and ease of programming and at the same time optimizes performance by allowing the overlap of computation and communication to tolerate the latency of remote data moves. In this paper we describe NNT and the implementation of the Well Posed Topographic model (WPT), a finite difference approximation weather prediction model. We present a qualitative study of the performance of the code on various multiprocessors and evaluate the effectiveness of NNT.

1. Introduction

Distributed memory multiprocessors with local address space are commonly used for high speed computing. In these machines data is shared among processors by sending and receiving messages containing that data. This movement of data is controlled explicitly by the programmer via send and receive instructions. The fact that the programmer has to explicitly control the placement and movement of data has been the main argument against local address space machines. Several software systems have been successfully developed to alleviate this problem. Linda [1] defines a "Tuple Space" that is globally accessible, and although it simplifies local address space programming, Linda compromises system performance for general applications. A tool that automatically transforms sequential Fortran 77 into code for local address space multiprocessors has been proposed [2], but its utility has only been proven for very regular program structures.

F.-X. Le Dimet (ed.), High Performance Computing in the Geosciences, 1–23.

Another approach to simplify the programming of local address space multiprocessors is the development of special purpose libraries. PDEQSOL [3] included library calls for the solution of partial differential equations. Several linear algebra algorithms [4] are being ported to distributed memory machines. These libraries execute complete mathematical functions with the necessary data manipulations. If the goal of the libraries is to isolate the programmer completely from complex data manipulations, their utility is limited to those algorithms that can be expressed in terms of the functions provided and to situations where data manipulation across processors occurs only in the respective algorithm function. For complex applications, like weather prediction models, this type of library will not suffice.

Our solution is to build libraries that are orthogonal to the mathematical functions executed on the local data. These libraries should provide calls for the placement and movement of data across processors. The library SILK [5] included procedures for data decomposition, I/O, remote data reference (including exchanges) , and global operations. Express [6] combined low level message passing and I/O routines with higher level calls for data decomposition, global operations, and two dimensional I/O. A high level library for grid oriented problems was proposed in [7]. It allows one to define a block-structured grid with boundaries between blocks, or overlap regions. The data in the overlap regions define the data that needs to be communicated through exchanges between processors.

The Nearest Neighbor Tool (NNT) is a library that isolates the user from all data manipulation procedures. The uniqueness of NNT is the programmer interface and its ability to increase performance by overlapping communication and computation. The programmers interface is simplified by concentrating on the functions necessary to solve Finite Difference Approximation (FDA) models. Through simple library calls, the user specifies at run time the way the data is to be decomposed among processors and the amount of data to be shared between them, or data boundaries. This information guides I/O, allocation, and exchange operations.

The overlap of computation and communication is crucial for the scalability of multiprocessor computers, since it allows the cost associated with remote data transfers to be hidden. NNT provides two library calls to overlap the exchange operations with the computation on local data: *NNTStartExchange* and NNT*EndExchange*.

NNT was coded using a layered approach and has been successfully ported to the Intel iPSC/860, Intel Paragon, TMC CM-5, nCUBE nCUBE2, KSR KSR1, and networks of workstations. Codes that are developed using NNT run on all these architectures without source code modifications.

In the following section we provide a description of NNT. In Section III we describe the implementation of the Well Posed Topographic (WPT) model. In Section IV we study the performance results of WPT running on a network of Sparc workstations, the Intel iPSC/860, the Intel Paragon, and the Kendall Square Research KSR1. We will use the results of the runs on the KSR1 to evaluate the cost of programming a global address space machine using message passing. Finally, in Section V, we describe the results of the work and mention future work that will be undertaken with NNT.

2. A General Description of NNT

We target data parallel applications, where parallelism is achieved by executing the same or different code on different data. In finite difference approximation weather models, for example, a set of partial differential equations is solved for each point on a space grid. The parallelism can be obtained from the degree of independence between the computations related to each of the grid points. In general, the operations that a local address space program of this type has to execute are the following :

a) Define local data space: data decomposition and memory allocation
b) Input and output operations
c) Computations upon local data
d) Communications of non-local data

NNT is targeted to the manipulation of data arrays. The data allocation and movement that are included in NNT are those found in FDA weather forecast models. NNT defines data boundaries between processors, or sections of arrays that are shared between neighboring processors. We present in the remainder of this section a description of the most important NNT calls, using the C binding version. For a formal, complete description of the calls refer to the NNT Users Manual [8].

2.1. DATA DECOMPOSITION

Each call to the decomposition function will return a decomposition structure that characterizes the way data is to be allocated, read, communicated (communication will occur by boundary exchanges), and written. The decomposition structure is an NNT type, named *nnt_decomp_struct*. A decomposition structure is associated with each array and used as input to NNT I/O and communication routines. The call to create a decomposition structure is *NNTDecomp*.

NNTDecomp*(DType, Boundary, GlobalSize, Thickness);*
DType: decomposition type (see Figure 2.1.1) (input)
Boundary: indication of periodic or non_periodic boundary conditions (input)
GlobalSize: indication of the size of the global array to be decomposed (input)
Thickness: indication of the thickness of each boundary (input)

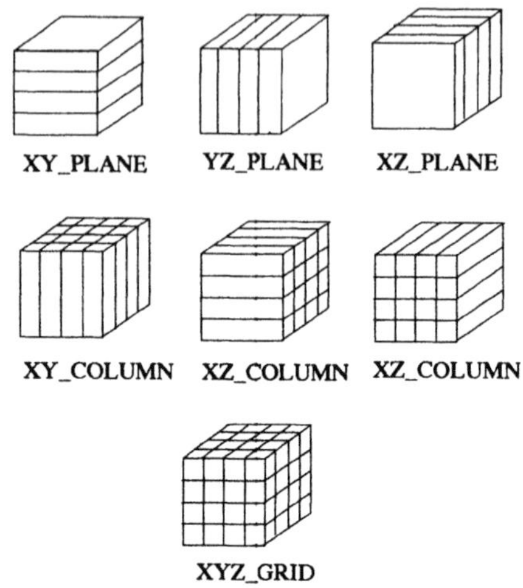

Figure 2.1.1. NNT decomposition types

2.2. MEMORY ALLOCATION

Each processor should execute forecast operations upon the arrays that are allocated and that correspond to the different variables in the atmosphere region that the processor is assigned to compute. These arrays contain interior data and boundary data. They also contain the data corresponding to the neighbors boundary regions. Memory is allocated in each processor to store the data local to the processor and the boundary data of the neighboring processors. Figure 2.2.1 shows the structure of a 2 dimensional data array allocated by NNT. The *local_boundaries* contain data that is within the region assigned to the processor. This data is sent to neighboring processors by boundary exchanges. The *non_local boundaries* contain the local boundary data of the neighboring nodes. This data is brought from the neighboring processors by boundary exchanges. A particular processor will use the non_local boundary data to compute the local boundary data. The *interior data* is only dependent on the local boundary data. The NNT call for memory allocation is *NNTAlloc*. *NNTAlloc* returns a pointer to the array that was allocated.

***NNTAlloc**(Dim, GSize, GDecomposed, Decomp, DataType);*
Dim: indication of the dimension of the array being allocated (input)
GSize: indication of the size of the array being allocated (input)
GDecomposed: indication of the dimensions to be decomposed (input)
Decomp: decomposition structure associated with the array to be allocated (input)
DataType: indication of the type of data contained in the array: int, float, ... (input)

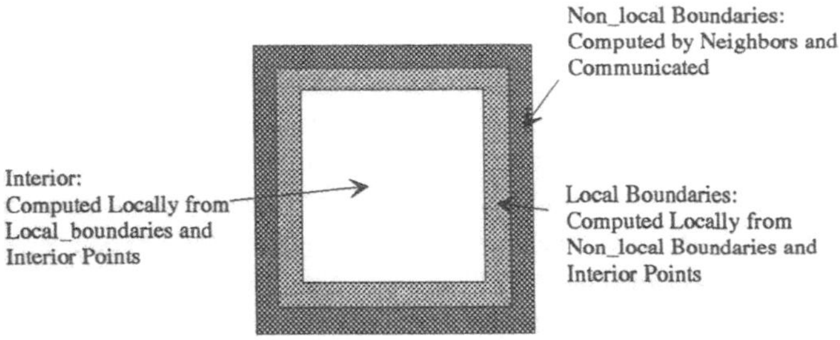

Figure 2.2.1. Local storage allocated by NNT for a 2 dimensional array that has been decomposed and contains boundaries is both dimensions.

2.3. INPUT AND OUTPUT OPERATIONS

NNT provides functions to read and write arrays from and to data files. ***NNTRead*** and ***NNTWrite*** are the I/O functions. The I/O operations read and write the boundary values. NNT I/O operations require each data array to be stored in a single data file, sequentially.

> ***NNTRead****(u, Decomp, File, DataType, Instance);*
> ***NNTWrite****(u, Decomp, File, DataType, Instance);*
> *u:* indication of the array to be read/written (input)
> *Decomp:* decomposition structure associated with the array (input)
> *File*: indication of the input/output file (input)
> *DataType:* indication of the type of data to be read/written: int, float, ...(input)
> *Instance*: indication of the instance of the array to be read/written (input)

The instance of the array is necessary to access a file with multiple copies of an array. This is common when a copy an array is written into a file at different simulation times. All arrays in a single file must have the same decomposition

2.4. EXCHANGE DEFINITION

Each exchange needs to be "defined" before it is carried out. The exchange definition function (*NNTDefineExch*) returns an exchange definition structure of type *nnt_exch_def*. This structure defines the gather/scatter procedures for the boundary exchanges and allocates buffers necessary for the message communication. Since all FDA weather prediction models use iterative algorithms, one exchange structure will be used for multiple calls of a particular boundary exchange.

> ***NNTDefineExch****(Decomp, DataType);*
> *Decomp:* decomposition structure associated with the array to be exchanged (input)
> *DataType:* indication of the type of data to be exchanged: int, float, ... (input)

2.5. COMMUNICATION OF NON-LOCAL DATA

Boundary exchange is the only data movement operation in NNT. NNT provides two methods for exchanging boundaries. One is a blocking exchange (*NNTExchange*) and the second is a non blocking exchange (*NNTStartExchange, NNTEndExchange*). A process continues execution after the NNTStartExchange call is issued and blocks execution at the *NNTEndExchange*. The exact behavior of *NNTStartExchange* depends on the low level routines available in the host system. The start and end exchange functions are used to overlap communication and computation.

NNTExchange*(u, Decomp, Tag, ExchDef);*
NNTStartExchange*(u, Decomp, Tag, ExchDef);*
NNTEndExchange*(u, Decomp, Tag, ExchDef);*

> *u:* indication of the array to be exchanged (input)
> *Decomp:* decomposition structure associated with the array to be exchanged (input)
> Tag: indication of the message tags to be used during the exchange (input)
> *ExchDef:* indication of exchange definition used for the exchange (input)

2.6. EXECUTION GUIDES

NNT provides execution guides to compute operations over the local data (*NNTComputeGuide*), the interior data and local boundaries (*NNTOverlapGuide*), and the non-local boundaries (*NNTOuterGuide*). These guides provide starting and ending indices for loops. *NNTOverlapGuide* provides starting and ending indices to guide the computation of the local boundaries independently from the interior, which is necessary to overlap computation and communication through the use of *NNTStartExchange* and *NNTEndExchange*. *NNTOuterGuide* provides the indices to guide computations over the non-local boundaries, also useful in certain occasions for codes that overlap communication and computation.

NNTComputeGuide*(Decomp, Start, End)*

> *Decomp:* decomposition structure associated with the array (input)
> *Start*: indices defining the beginning of the local data (output)
> *End*: indices defining the end of the local data (output)

NNTOverlapGuide*(Decomp, NBoundary, BStart, BEnd, IStart, IEnd)* *Decomp:* decomposition structure associated with the array (input) *NBoundary*: number of boundaries (output) *BStart*: indices defining the beginning of the local boundaries (output) *BEnd*: indices defining the end of the local boundaries (output) *IStart*: indices defining the beginning of the interior (output) *IEnd*: indices defining the end of the interior (output)

NNTOuterGuide*(Decomp, NBoundary, Start, End)* *Decomp:* decomposition structure associated with the array (input) *NBoundary*: number of boundaries (output) *Start*: indices defining the beginning of the non-local boundaries (output) *End*: indices defining the end of the non-local boundaries (output)

3. The Implementation of the Well Posed Topographic Model

3.1. THE WELL POSED TOPOGRAPHIC MODEL

The Well Posed Topographical (WPT) model is a non-hydrostatic, compressible meteorological model based on the "approximate system" of Browning and Kreiss [9] plus the incorporation of topography as described in Browning and MacDonald [10]. The first version of the model was coded using F77 and it was later parallelized by Baillie and MacDonald [11] to run on a Kendall Square Research KSR1 multiprocessor. We based our parallelization effort on the KSR parallel code. In particular, we adopted the code restructuring that was proposed to eliminate data dependencies that prevent parallelism. The first version of WPT forecast five (5) variables: u (west wind), v (south wind), w (vertical wind), p (pressure), and θ (potential temperature). A temporary variable, wp, is also carried throughout the forecast. "New" values for these variables are calculated using fourth-order centered space differencing in both the horizontal and vertical directions. The code can be divided in two main loops, the Forcing Loop that calculates forcing functions from the "old" values from advection, dynamic, and topographic equations, and the AB Loop that calculates "new" values using the third-order Adams-Bashforth scheme:

Forcing Loop

input:	u_{old}, v_{old}, w_{old}, wp_{old}, p_{old}, t_{old}
output:	fu_{new}, fv_{new}, fw_{new}, fp_{new}, ft_{new}

AB Loop

input:	fu_{new}, fv_{new}, fw_{new}, fp_{new}, ft_{new}
	fu_{old}, fv_{old}, fw_{old}, fp_{old}, ft_{old}
	fu_{very_old}, fv_{very_old}, fw_{very_old}, fp_{very_old}, ft_{very_old}
output:	u_{new}, v_{new}, w_{new}, wp_{new}, p_{new}, t_{new}

We will leave further details of the model to references [9] and [10]. The important characteristic to understand is that both loops are parallelizable since they do not contain data dependencies between loop iterations. The Forcing Loop has to be executed before the AB loop for a particular time iteration. The AB loop contains data dependencies across three (3) time iterations: new, old, very_old[11].

3.2. NNT IMPLEMENTATION

Since there are no data dependencies across iterations of the Forcing loop and the AB loop, we can parallelize WPT by mapping each processor to a particular atmospheric volume. This mapping is done by *NNTDecomp*. The general flow of the program, with the main NNT calls, is shown in Figure 3.2.1. We have written three versions of the parallel program, which differ only in the forecast loop. The differences between these versions will be shown later. The NNT call are shown in italic.

The data decomposition strategy is selected by the user at run time, and the decomposition parameters are defined by *NNTDecomp*. Fourteen (14) arrays have to be allocated (*NNTAlloc*) according to the chosen decomposition, fourteen (14) arrays have to be read (*NNTRead*) from initial data files, five (5) arrays have to be written (*NNTWrite*) to output final values for the winds, temperature and pressure. In the forecast iteration loop, after the forecast functions have been executed, data is exchange using *NNTExchange* to update the boundaries in each of the processors.

The goal of NNT is to isolate the parallel programming issues from the computation of the forecast functions. The original sequential code was written in Fortran 77, and any parallelization work was to avoid considerable changes to this code. To this effect, we wrote a "Parallel Driver" that performed all the operations related to parallel programming (decomposition, allocation, I/O, and communication) and that called the Fortran 77 forecast code as a subroutine. This subroutine was obtained by generalizing the original sequential code to operate in a atmospheric volume defined by its dimensions and its position in the total volume of forecast. The changes to the sequential code were minimal: about 25 lines of Fortran 77 to set up the minimum and maximum loop indices values for the Forcing and the AB loops.

The Parallel Driver contains all the NNT calls. It was written in C, but a Fortran interface containing a subset of the NNT calls is also available. The only NNT calls that were not implemented in Fortran are the memory allocation and memory deallocation routines, since standard Fortran 77 does not allow for dynamic memory allocation. *NNTDecomp* is called for non-periodic boundaries and boundary thickness of two (the code performs fourth-order differencing). The user selects at run time the size of the total atmospheric volume, the number of processors, and the decomposition type (by columns or rows, 2D mesh, 3D grid, etc.). The forecast is performed by iterating over the execution of the Forecast Subroutine. The total number of instructions in the iteration loop is 32. Only 6 are NNT calls: 6 array exchanges.

```
Parallel WPT
        Decomp = NNTDecomp( decomposition_type, ...)
        array1 = NNTAlloc(..., Decomp, ...)
                        .
                        .
        array14 = NNTAlloc(..., Decomp, ...)
        NNTRead(array1, Decomp, InputFile1, NNTFloat, ...)
                        .
                        .
        NNTRead(array14, Decomp, InputFile14, NNTFloat, ...)
        ExchDef1 = NNTDefineExch(Decomp, NNTFloat)
                        .
                        .
        ExchDefN = NNTDefineExch(Decomp, NNTFloat)
        NNTComputeGuide(Decomp, ...)      (for version 1)
        NNTOverlapGuide(Decomp, ...)      (for version 2 and 3)
        NNTOuterGuide(Decomp, ...)        (for version 2 and 3)
        Start Iteration
                Compute Forcing and AB loops and Exchange (3 versions)
        Continue Iterating until done
        NNTWrite(array1, Decomp, OutputFile1, NNTFloat, ...)
                        .
                        .
        NNTWrite(array5, Decomp, OutputFile5, NNTFloat, ...)
End program
```

Figure 3.2.1. A high level description of the implementation of the Well Posed Topographic Model. N arrays need to be exchanged. N is 5 for Versions 2 and 3, and 6 for Version 1.

The structure of the iteration loop, for a non-optimized version, is the following:

<u>WPT iteration. Version 1</u>

Iterate

Call Forcing and AB loops for Local Data

NNTExchange(u, Decomp1, ...)

NNTExchange(v, Decomp1, ...)

NNTExchange(w Decomp1, ...)

NNTExchange(wp, Decomp1, ...)

NNTExchange(p, Decomp1, ...)

NNTExchange(t, Decomp1, ...)
Continue Iterating

To minimize the amount of data communicated, the Forcing and the AB loops are split into different subroutines. The Forecast Subroutine is called as before, but only five (5) forcing variables are exchanged. After the exchange, the AB loop is executed for the non_local boundaries to obtain the new values for the dependent variables from the forcing variables exchanged; instead of communicating the original variables (*u*, *v*, *w*, *p*, *t*, *wp*), we communicate the forcing variables and compute the original variables on the non_local boundaries:

<u>WPT iteration. Version 2</u>
Iterate
Call Forcing and AB loops for local boundaries
NNTExchange(fu)
NNTExchange(fv)
NNTExchange(fw)
NNTExchange(fp)
NNTExchange(ft)
Call AB loop for non_local boundaries
Continue Iterating

The communication is reduced by 16.6 % and the computation is increased by the size of the boundaries. Depending on the size of the problem and number of processors, and on the ratio of processor speed to network speed, the first or the second approach should produce better performance results.

The final version that we present is one that overlaps communication with computation to hide communication costs. The Forcing loop is executed for the local boundary values and then the forcing variables exchange is started by *NNTStartExchange*. This exchange is overlapped with the computation of the Forcing loop for the interior points and the computation of the AB loop for the interior and local boundary points. The exchange is ended by NNTEndExchange-exchange. Then the AB loop is executed for the non_local boundary. The iteration loop for the overlapped version is the following:

<u>WPT iteration. Version 3</u>
Iterate
Call Forcing loop for local boundaries
NNTStartExchange(fu)
NNTStartExchange(fv)
NNTStartExchange(fw)
NNTStartExchange(fp)
NNTStartExchange(ft)
Call Forcing loop for interior

Call AB loop for local data
NNTEndExchange(fu)
NNTEndExchange(fv)
NNTEndExchange(fw)
NNTEndExchange(fp)
NNTEndExchange(ft)
Call AB loop for non_local boundaries
Continue Iterating

4. Performance Results and Qualitative Considerations

To study the performance of Well Posed Topographic model coded in NNT we ran several experiments on a Ethernet network of Sparc2 workstations, the Intel iPSC/860, the Intel Paragon, and the Kendall Square Research KSR1 multiprocessor. We use the results from the workstations network and the iPSC/860 to study the effectiveness of NNT to overlap computation and communication. We use a 400 node Intel Paragon to show the performance of the code in highly parallel machines, and discuss the performance of the system. Finally, we use the results from the KSR1 to qualify the cost of portability. The main cost of programming in message passing for portability is the overhead of running such codes in a shared memory environment, since software controlled data movement is used on an architecture capable of controlling the movement of data in hardware.

4.1. OVERLAPPING COMPUTATION AND COMMUNICATION

Figure 4.1.1 shows the performance of the three versions of WPT running on an Ethernet network of Sparc2 workstations for a problem size of 130x98x31. The performance is measured in millions of single precision floating point operations per second, or single precision MFLOPS. We can see that the scalability of the three versions is approximately the same until four processors, achieving a speedup of about 3. For five and six processors, however, each program behaves differently. Version 3 of the code, which overlaps communication and computation, achieves the best scalability, with about a 3.8 speedup for six processors. Versions 1 and 2 achieve speedups of about 2.1. The performance of the system decreases after 4 processors for version 1 and 2.

There are two effects that contribute to the performance behavior of the applications. The first one is related to the speed of the interconnection network, the second related to the single processor performance. The problem size for all the runs was 130x98x31. The data volume was partitioned using an NNT XY_Column decomposition for 4 and 6 processors. For 1, 2, 3, and 5 processors an NNT XZ_Plane decomposition was used (see Figure 2.1.1). In general, as we increase the number of processors in the system, the overhead (latency) of boundary exchanges increases because of increased contention for the network, which is very high for Ethernet (bus network). Version 2 of the algorithm scales better that Version 1 since there is a reduction of 16.6% on the amount of data communicated (see Section 3.2). Version 3 of the algorithm scales considerably better than the two other versions since the data communication is overlapped with computation. This large performance improvement is expected given the small bandwidth available in the Ethernet network. However, we can note that for four processors the performance of Version 1 is higher than Version 2 or 3. The reason is the behavior of the single node performance.

The code that computed the forecast function (the Forcing and the AB loop) was optimized to run faster on processors where data locality is an issue. The optimization was proposed by Bruce Greer, of Intel Supercomputers. It consisted of increasing the locality of reference of the code by looping in the X direction around each instruction that calculated a finite difference type operation. The code produced approximately a 2 fold increase in performance in the Sparc2 and a 3 fold increase in performance in the i860 microprocessor (node for the Intel Supercomputers). However, due to the overhead in executing a loop operation, a similar effect to that found in vector computers was observed, where the efficiency of the computation decreases as the vector length is decreased bellow a certain value [4]. Figure 4.1.2 shows the single processor performance for WPT for three multiprocessors as the size in the X dimension in increased (East-West). We can notice that for small sizes in the X axis, the single node performance is very low, and increases rapidly until it reaches a stable point.

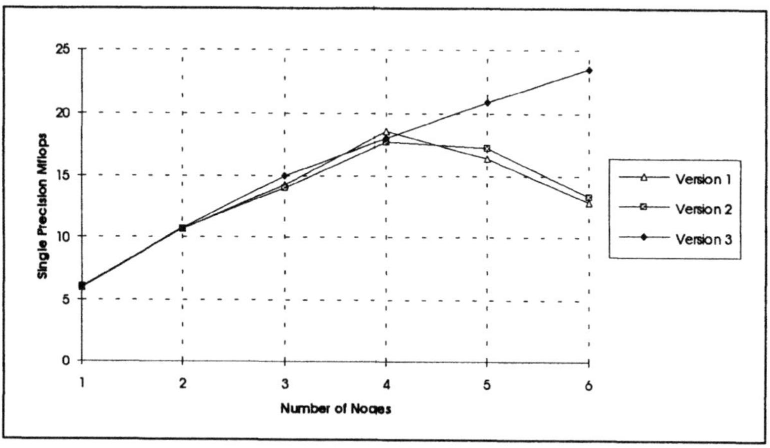

Figure 4.1.1. Total performance, in MFLOPS, for WPT running on a network of Sparc2 workstations connected by an Ethernet network. Problem size: 130x98x31. The graph shows the results for the three versions of the code described in Section 3.2. Results are from *single precision* operations.

The behavior of the single processor performance shown in Figure 4.1.2 causes versions 2 and 3 of WPT to run slower than Version 1 for 4 processors. Versions 2 and 3 compute data on the local and non-local boundaries (see Section 3.2). For the XY column decomposition selected for 4 and 6 processors, there are boundaries of thickness=2 in the

X direction, and therefore, while executing the AB or Forcing loops on the local or non-local boundaries, a Sparc2 processor will be executing at approximately 2 MFLOPS. This was verified by measuring the processor speed independently from the communications overhead. For 6 processors, although there are boundaries in the X direction, the communication overhead was so large that decreasing the amount of communication and overlapping communication and computation produced considerable performance increase.

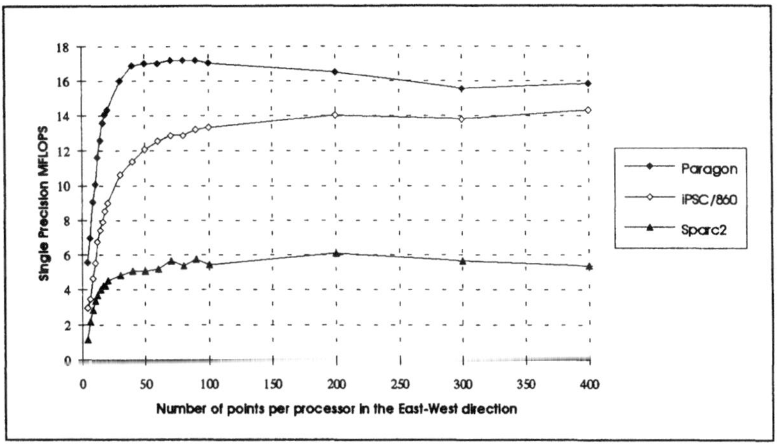

Figure 4.1.2. Single node performance for WPT running on a single node of the Paragon, iPSC/860, and Sparc2. The performance is plotted for variations of size of the data in the East-West direction.

The large increase in performance by communication and computation overlap was possible because there was a large network effect to hide. In machines were there is a better balance between processor and network speed, this performance improvement is more difficult to obtain. In particular, fully asynchronous, non-blocking communication becomes necessary. Figure 4.1.3 shows the performance of versions 1 and 3 of the algorithm on the iPSC/860. Our libraries only supported blocking communications, that although not synchronous, prevent the efficient overlap of boundary exchange and computations. We can see in Figure 4.1.3 that the performance increase in the iPSC/860 was considerably smaller than the one for the Sparc2 network. The reason is that the iPSC/860 is a better balanced machine compared to a workstation network, the ratio of network speed to processor speed is higher. We also see that for 12 processors and a small data size, the overlapped version performs worse than Version 1, since after that point the local computation is not enough to overlap the communication and the overhead necessary for the overlap becomes a limiting factor. The results for the Version 2 are not shown since this version always ran slower than Version 1. The reason is that, for the iPSC/860, the reduced communication in Version 2 is small compared to the extra code that needs to be executed.

The goal of this section was to study the effectiveness of NNT to overlap communication and computation and increase system performance. We showed how the performance of multicomputers with a slow network can be substantially increased by using libraries that overlap boundary exchanges and interior computations in FDA weather prediction models. The benefits of the overlap is reduced for two reasons: the behavior of the single node performance and the overhead of communication that is not fully asynchronous.

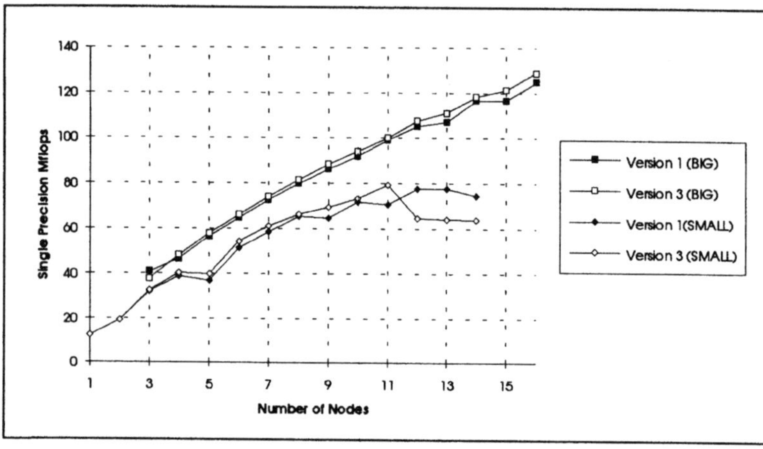

Figure 4.1.3. Performance of Versions 1 and 3 of WPT running on the Intel iPSC/860. The BIG and SMALL problem sizes are 132x98x31 and 60x60x32 respectively. Performance is measured in *single precision*.

To improve the benefits of data movement overlap the *code optimizations for single node performance have to be designed so that the single node performance is also high during boundary computations.* For systems with good balance of computation and communication speed, *fully asynchronous communications is necessary.* We ran experiments on the Intel Paragon and there was no performance improvement by overlapping the boundary exchanges.

4.2. PERFORMANCE ON A 400 NODE PARAGON

In the next experiment we studied the scalability of codes by running the Version 1 of WPT on an Intel Paragon with 400 nodes. As we mentioned before, the overlapped version of WPT did not improve performance since our implementation of NNTStartExchange and NNTEndExchange did not use fully asynchronous sends and receives. The code was run with a problem size of 400x300x50. Figure 4.2.1 shows the total system performance and the single node performance (total performance divided by the number of processors) as the number of processors is incremented from 100 to 400. The single node performance is plotted with empty objects and the total system performance with filled ones.

We discuss three program characteristics that contribute to the performance behavior observed in Figure 4.2.1; the X direction vector length, the locality of data reference, and the locality of communication.

For the runs shown in Figure 4.2.1 we selected a XY_Column NNT decomposition (see Figure 2.1.1). As the number of processors is increased, the length of the X direction per processor is decreased. Due to our sequential code optimization this causes the processor to compute with less efficiency, as shown in Figure 4.1.2. This causes the single node performance to decrease with the number of processors, as can be seen in Figure 4.2.1 for more than 144 processors. At 144 processors the length of the vectors in the X direction is about 34 (12 processors in the X direction), and from Figure 4.1.2 we can see that below this point the single node performance decreases significantly.

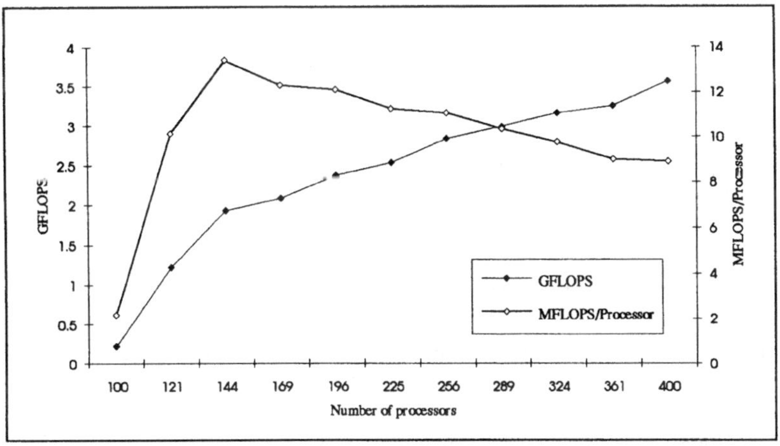

Figure 4.2.1. Total machine performance and single node performance for WPT running on the Intel Paragon. The data size is 400x300x50. The dark objects indicate total system GFLOPS and the light objects indicate MFLOPS per processor

The rise in the curve of single node performance in Figure 4.2.1 is due to the reduction in the working set for each processor. For 100 and 122 processors the local data size does not fit in memory and paging is necessary between main memory and disk. As we increase the number of processors the working set per processor decreases lowering the demands on the memory hierarchy.

Observing Figures 4.2.1 and 4.1.2 we can note that the loss in single node performance is not sufficient to explain the decrease in system efficiency as we increase the number of processors. In Figure 4.2.1 the single node performance for a X dimension vector length of 34 is about 13 MFLOPS, where our experiment results presented in Figure 4.1.2 show that it should be approximately 16 MFLOPS. We believe that the contention for network resources is responsible for the rest of the performance loss; as the number of processors in increased, the contention increases as well. However, since our code only executes nearest neighbor communication, and the Paragon has a 2 dimensional mesh network, it would seem natural that the contention for network paths does not increase as the number of processors increase. This, however, assumes an ideal mapping of virtual processors to real processors. To prove that the contention for network resources increases with the number of processors we ran an experiment that kept the problem size per processor constant as the number of processors participating in the computation increased. In this way, the single processor speed is kept constant (constant X direction vector length and data locality characteristics), and the only effect that is visible is the interconnection network. Figure 4.2.2 shows the execution time of an iteration of the Version 1 of WPT, from 1 to 400 processors. The straight line indicates the expected execution time of the system if no communication were present.

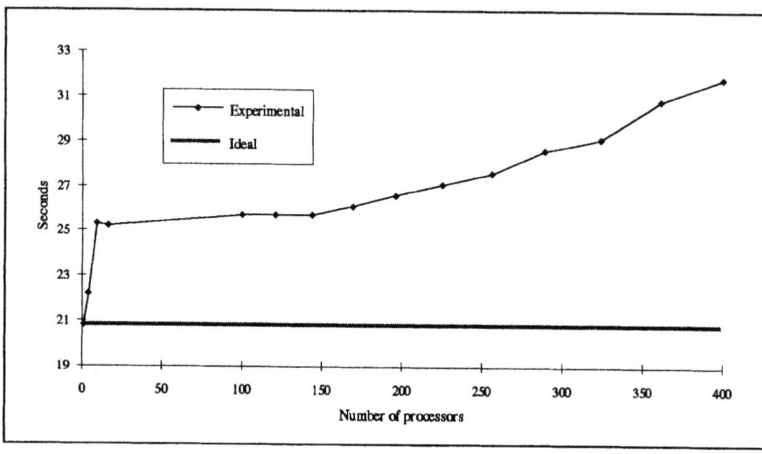

Figure 4.2.2. Execution time for an iteration of WPT as the problem size per processor is kept constant as the number of processors participating in the computation increases. The single node domain is 20x15x50. The figure shows experimental and ideal execution time, where the cost of communication is zero.

The experimental results show the initial rise in execution time due to the increase in communication per processor. After 16 processors the execution time is dominated by processors that have to exchange boundaries with four other processors. For more than 16 processors the average amount of data communicated is constant for interior processors. We see, however, that the execution time increases with the number of processors. Since the number of computations and the computational speed of each processor is constant, and the amount of communication is constant, we can conclude that the increase in execution time is due to the decrease of communication speed. This is caused by a latency increase caused by an increase contention for network resources. The reason is that the nearest neighbor communication in NNTs virtual array of processors does not translate into nearest neighbor communication in the physical processor array. Although the Intel Paragon is a 2 dimensional mesh, and the processor numbers selected are perfect powers of two, the physical processor array scheduled for the computation might not be a perfect square, since it depends on the availability of nodes and shape of the "partition" assigned by the Paragon system. In this way a 144 processor might be taken by NNT as a 12x12 processor array, but physically, a 16x9 processor array is used. Boundary exchanges between virtual processors might involve multiple path communications between physical processors.

To improve the performance of the code two issues are fundamental. The first one is to *select a single node performance optimization strategy that does not suffer from degradation as the number of processors increases*, as characterized by Figure 4.1.2. In our case the problem was that the decomposition strategy subdivided the X dimension across processors, decreasing the single node performance. A two dimensional decomposition is necessary in order to reduce the ratio of communication to computation. We could not use a YZ Column decomposition because the code presented data dependencies in the Z axis. The other alternative is to optimize by increasing data reusability in the Z dimension, instead of the X dimension. However, preliminary analysis showed that the single node performance improvement is the highest by optimizing in the X dimension.

The second consideration in improving the total system performance is to *optimize the decomposition for the physical processor distribution*. For the Paragon, and a two dimensional decomposition, this means a simple selection of the size of each dimension. NNT currently uses the most square virtual processor array, but to handle situations as the one described above, the user should be allowed to specify the shape of the virtual processor array so as to match the available partition on the system. This should be defined at run time.

4.3. THE COST OF PORTABILITY

NNT is built upon message passing. Message passing was selected because it is currently the lowest level common denominator between parallel architectures. Message passing can be implemented in global address space architectures, and it is practically available in all the global address space multiprocessors in the market. However, there is a cost associated with programming global address space machines using message passing. Global address space computers are build so that the hardware controls the movement of data, generally on demand. In local address space architectures the programmer, by using send and receive instructions, controls the movement of data. The movement of data in local address space multiprocessors is initiated by the producer of the data (with the exception of new active message implementations). The cost of using message passing in global address space multiprocessors is mainly the penalty for software overhead and extra memory copying that occurs.

To investigate the cost of programming in message passing a global address space multiprocessor we ran the Version 1 of WPT on the KSR using or NNT implementation, and the implementation of NNT that was developed by Baillie and MacDonald [11] using the KSR tiling directives. NNT used the TCGMSG message passing libraries available on the KSR. For the NNT/WPT runs we chose a one dimensional decomposition that matched the tiling strategy for the native KSR implementation. Figure 4.3.1 presents the results of the runs for a problem size of 130x90x30. Note that the results are double precision, and hence should not be compared with the performance results reported in the previous sections. The important issue to notice is the difference in performance between the message passing (NNT) and shared access implementations.

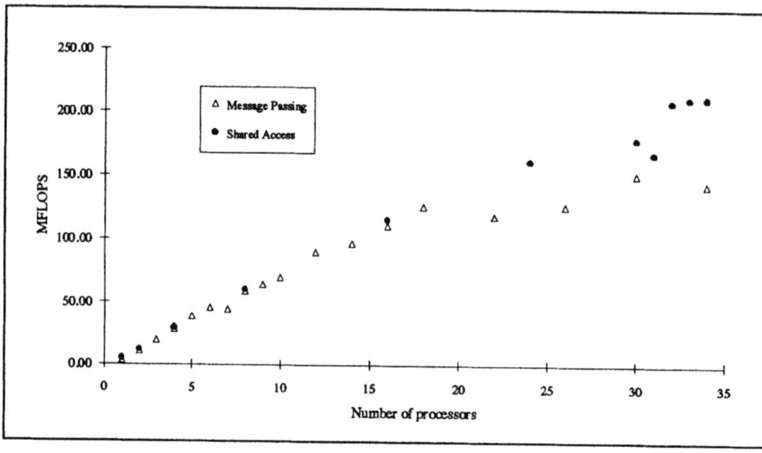

Figure 4.3.1. Total machine performance, in MFLOPS, for a global address space and a local address space version of WPT. The data size is 130x90x30.

In Figure 4.3.1 we see that the performance for the two codes is practically the same for small number of processors. However, after 20 processors the performance of the code that uses NNT increases slower than that of the shared access implementation. This is due to the decrease in the granularity of the computation. For smaller number of processors the communication or data movement time is very small compared to the computation time. As we increase the number of processors the ratio of data movement to computation increases, and with it, the overhead caused by software intervention in message passing. Figure 4.3.2 shows the system performance for the NNT and Tiled versions of WPT for a smaller problem size. Note that the performance of the message passing version starts to deviate from the shared access version after two processors.

Besides the overhead from the message passing calls, the system incurs unnecessary data copying. First the data has to be gathered into a buffer and then scattered to the appropriate addresses in memory. The KSR1 is a Cache Only Memory Architecture, or COMA [12]. It has a local memory associated with each processor and data sharing is accomplished by the automatic movement of data between processor's memories. The address space is dynamically mapped to local memories, so in effect, an address location "moves" to the cache of the processor accessing a data item. The movement of data is managed by data coherence protocol implemented in hardware. Figure 4.4.2 presents a diagram of the movement of data necessary to transmit a boundary between two processors.

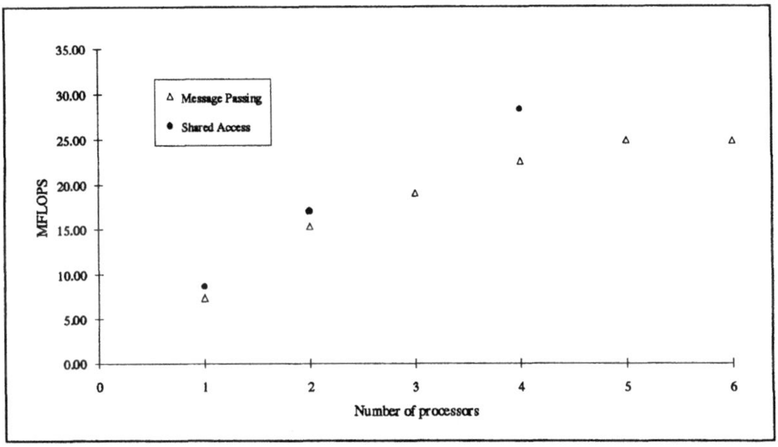

Figure 4.3.2. Total machine performance, in MFLOPS, for a global address space and a local address space version of WPT. The data size is 31x24x21.

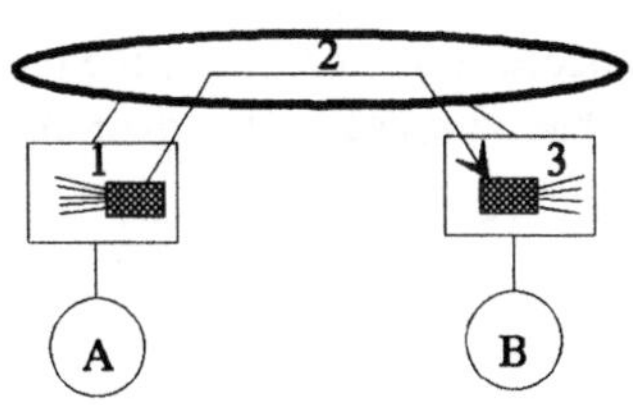

1) A gathers data from memory locations into communication buffer
2) B accesses communication buffer, transfering data from A to B
3) B scathers data from communication buffer to memory locations

Figure 4.3.3.a. Data movement on the KSR1 using message passing. When processor A executes a send the required data is gathered into a communication buffer. When processor B executes a receive the communication buffer is transferred from A to B and the data is scattered from the communication buffer into the respective memory locations.

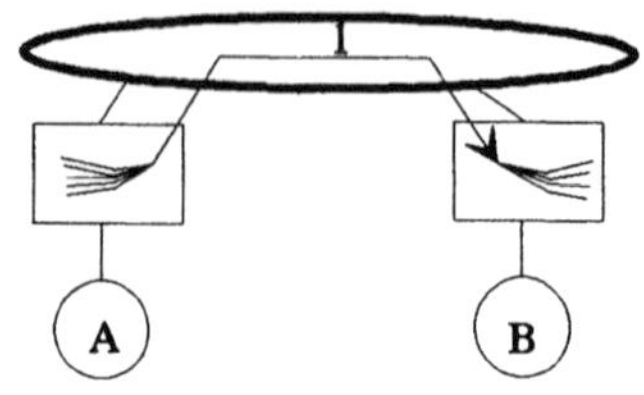

Figure 4.3.3.b. Data movement on the KSR1 using shared access. Processor B accesses the data in A causing it to be transferred to its local memory (cache).

The extra data movement due to message passing are steps 1 and 3 of Figure 4.3.3.a. This data movement is inside the processor's cell and involves data transfers between memory locations of the processor's cache. A careful analysis can determine the cost associated with this extra data movement and the cost associated with the software overhead. To try to quantify both we ran an experiment where a vector of 5000 64-bit floating point numbers were transferred between two processors A and B. The vector is first initialized in A. In the native KSR implementation processor B accesses the vector causing the data to be transferred from A's memory to B's memory (Step 1 in Figure 4.3.3.b). In the message passing implementation A executes a send to B and B a receive from A. The data in the vector is gathered and scattered as indicated in Figure 4.3.3.a. Figure 4.3.4 shows the total time necessary to complete the data transfer. The data transfer controlled by message passing is about 3 times slower than that controlled by the hardware.

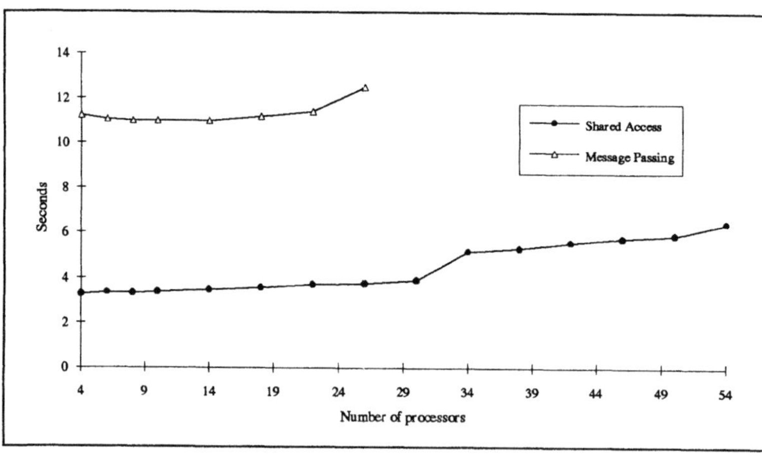

Figure 4.3.4. Time to read 5000 floating point elements (64-bits) using shared data access and message passing on the KSR1 multiprocessor.

The difference in transmission times shown in Figure 4.3.4 is the basic measure of the cost of programming the KSR1 using message passing. The degree in which this affects the overall execution time depends on the ratio of computation to communication, as seen in Figures 4.3.1 and 4.3.2.

There is another effect that makes the data movement by shared access more efficient than that of message passing. When executing a receive instruction the data is transmitted through the network all "at once". The processor accesses the transmission buffer without interruption causing a burst of traffic in the network. On the other hand, the accesses to shared memory locations move data on demand, so the individual data items that form the boundaries move through the interconnection network at the time that they are accessed. Since typical codes fetch data and compute on them before fetching more data, the data transmissions will be separated by periods of execution, lowering the requirements on the network when compared to message passing.

5. Conclusions

We have shown that it is possible to build a high level library to simplify the programming of FDA weather prediction models on local address space multiprocessors. The programming interface is very simple and provides functions for problem decomposition, array allocation, array read and write, and boundary exchanges. By using a layered approach, we keep constant the programming interface regardless of the underlying architecture. Programs developed in NNT will run on a variety of machines without any source modification, currently, the nCUBE2, the Intel iPSC/860 and Paragon, network of workstations, the Thinking Machines CM-5, and the KSR1. We are currently programming the Eta forecast model using NNT. Eta is a complex weather prediction model that is currently run at the National Meteorological Center.

There are several lessons that were learned during the preliminary phases of experimentation with NNT and WPT. In first place it is important to keep in mind that program optimizations targeted to improve single node performance may impact heavily the behavior under parallel execution. We discussed, for our example, that the effectiveness of overlapping the boundary exchange with local computation was reduced because the single node performance dropped significantly during the boundary computation. Our program optimizations for the i860 had similar conditions as those for vector machines, where the efficiency of the computation is small for small vector lengths. If the decomposition strategy subdivides the space in the "vectorized" dimension, the single node performance decreases with the number of processors. The second lesson learned is that although FDA models have very good communication locality (nearest neighbors) it is essential to optimize the mapping of virtual processors to real processors. This optimization should be done by the high level tool, such as NNT, at run time. In the Paragon, for example, the dimensions of the virtual array should match that of the partition allocated to the user.

The NNT calls for overlapping the data exchange with local computation are effective for multiprocessors with high communication latency. On an Ethernet network with six Sparc2 workstations, for example, the performance was increased by more than 30% by using the NNTStartEnchange and NNTEndExchange calls. To make data overlap more effective on multiprocessors with faster networks, the current version of NNT has to be upgraded to perform fully asynchronous communication.

WPT was run on the KSR1 global address space multiprocessor and the cost associated with message passing programming was evaluated. The results show that for problems with large granularity the performance of message passing programs is similar to that of programs written with the native KSR tiling directives.

Acknowledgment.

The authors of this paper will like to thank Clive Baillie and George Carr for reviewing this paper and creating interesting discussions, Bruce Greer for optimizing the single processor version of WPT, Jacques Middlecoff for his assistance with the KSR1, Sandia National Laboratories and San Diego Supercomputer Center for providing access to their Paragon multiprocessors, and Dave Gale for making the access to the 400 node Paragon at Sandia possible with such a short notice.

References

[1] N. Carriero, and D. Gelernter, "Linda in Context," Comm. of the ACM. v.32, no.4, 444-458. 1989.

[2] M. O'Keefe, T. Parr, B. Edgar, S. Anderson, P. Woodward, and H. Dietz, "The Fortran-P Translator: Automatic Translation of Fortran 77 Programs for Massively Parallel Processors.

[3] H. Hirayama et. al. "Solution Functions of PDEQSOL (Partial Differential EQuation SOlver Language) for Fluid Problems," Proceedings of Supercomputing '91, ACM Press, Albuquerque, New Mexico, November 1991.

[4] J. Dongarra, I. Duff, D. Sorensen, and H. van der Vorst, "Solving Linear Systems on Vector and Shared Memory Computers," SIAM, Philadelphia, Pennsylvania, 1991.

[5] S. Phillips, R. Willeman, and R. Fargason, "Integrated Tools for a Mathematics Library on Distributed Memory Systems," Proceeding of the 4th Conference on Hypercubes, Concurrent Computers, and Applications, Monterey, California, March 1989.

[6] Parasoft Corporation, Express C Reference Guide, Parasoft Corporation, 1990.

[7] R. Hempel, and H. Ritzdorf, "The GMD Communications Library for Grid-Oriented Problems," Technical Report 589, GMD, Sankt Augustin, Germany, 1991.

[8] B. Rodriguez, L. Hart, T. Henderson, "NNT 1.0 User's Guide," NOAA Technical Memorandum in preparation.

[9] G. L. Browning and H. O. Kreiss, "Scaling and computation of smooth atmospheric motions", Tellus 38A, 1986.

[10] G. L. Browning and A. E. MacDonald, "Incorporating topography into the approximate systems", submitted to Dynamics of Atmosphere and Oceans, 1992.

[11] C. F. Baillie and A. E. MacDonald, "Porting the Well-Posed Topographical Meteorological Model to the KSR Parallel Supercomputer," Proceedings of the Fifth Workshop on the Use of Parallel Processors in Meteorology, Reading, England, November, 1992.

[12] Kendall Square Research Corporation, "KSR1 Principles of Operation," 1991.

THE NEW CANADIAN VARIABLE-RESOLUTION MODEL: a universal tool to study atmospheric phenomena on a broad temporal and spatial scale

SYLVIE GRAVEL, JEAN CÔTÉ, ANDREW STANIFORTH, MICHEL ROCH, AND ALAIN PATOINE
Recherche en Prévision Numérique,
Environnement Canada
2121, Route Transcanadienne,
Québec, Canada
H9P 1J3

1. Introduction

1.1. PRESENT DAY FORECAST MODELS AT CMC

For the past seven years, the Canadian Meteorological Center (CMC) has had two operating forecast models: a global model to produce long term forecasts, and a regional model, to produce *quickly* regional forecasts.

The characteristics of the two models are the following:

The global model uses a spectral representation in the horizontal with a T119 resolution. In the vertical, it uses a finite-element representation. The model is semi-implicit, semi-Lagrangian, with a three time-level discretization. It produces ten day forecasts.

The Finite Element Regional (FER) model, as its name indicates, uses finite-element representation in the three directions. It integrates the primitive equations on a polar stereographic projection of the earth true at 60°N, using a cartesian variable grid.

Figure 1 illustrates the operational grid of the FER model: 39% of the grid points are on the high-resolution subdomain where the resolution is 100km. Outside this area, the distance between two adjoining grid points increases steadily by 10%. A higher resolution FER is presently tested in parallel runs at CMC. This model has a resolution of 50km on the same region of interest, and in this case 50% of the points cover the area.

The FER uses a three time-level semi-implicit, semi-Lagrangian scheme. The role of the FER is to produce quickly -- within 80 minutes after the synoptic observation time -- a 48h high-resolution forecast for Canada. It can achieve this goal since by contrast with the global model which requires a more complete set of data as initial condition, the FER can do with only hemispheric data.

1.2. THE COMPUTING CAPABILITES OF CMC

The models are run on CMC's supercomputer: a NEC SX3-44. This new machine has replaced a CRAY X-MP 4/32EA in December 1992. The increase in computing power and available memory has allowed the Canadian Meteorological Centre to increase the resolution of the spectral model, and soon that of the FER model.

F.-X. Le Dimet (ed.), High Performance Computing in the Geosciences, 25–32.

The NEC has a relatively conventional TCMP architecture (Tightly Coupled MultiProcessors): it has four processors, each with four 4-pipe functional units. There is one gigabite of central memory, and four gigabites of auxiliary memory.

1.3. THE NEW GLOBAL MODEL

The two operational models complement each other well. The regional model provides a high-resolution forecasts quickly, but is limited to producing 48h forecasts -- beyond that the rigid walls at the equator, and the insuffisant resolution in the area surrounding the region of interest introduce unacceptable errors. The global model is perfect for long integrations because of its uniform resolution, but needs a more complete set of data at initial time, and therefore can not be run quickly after synoptic time. Neither can it be run at very high resolution because of the limited computing power and memory.

The maintenance of two distinct operational models however, is costly: there are two distinct working groups, each having to look after, and debug impressive amount of computer code. The desire to have a unique model to perform both tasks of producing a short-range regional, and a medium-range global forecast and analysis is the incentive for the development of the new Canadian model.

Here are the main characteristics of the new model:

To achieve its double goal the model relies on a horizontally varying mesh which allows a focusing of resolution over an area of interest. The proposed operational strategy is to run the model in a variable-resolution regional configuration with a focusing over Canada to produce detailed 48h forecasts, and rerun the model in a uniform-resolution global configuration, to produce medium-range forecasts. The grid, with a resolution of approximately 50km on the area of interest is illustrated in Fig. 2. Note that the poles of the grid and the geographical poles do not coincide.

The construction of the grid is as follow: the size in latitude and longitude of the area of interest is determined, as well as the desired resolution. The decrease in resolution (usually around 10%) can either be specified or left to the grid generator. The grid, with these specifications is build around the equator so as to obtain grid points that are as equidistant as possible. Once the grid is generated it is rotated -- the geographical poles than no longer coincide with the grid's poles -- to bring the high-resolution area above the geographical area of interest. The use of a rotated grid leaves the governing equations completely unchanged.

The model uses a Galerkin method with a finite-element representation for the horizontal and vertical structure. The finite elements offer the flexibility required by the variable resolution grid.

The time discretization is done using a two-time-level semi-implicit semi-Lagrangian scheme. Present day operational models that are semi-Lagrangian use three-time-level discretizations, but a two-time-level formulation is potentially twice as efficient, since for the same level of time-truncation error, it allows timesteps twice as long as a three-time-level scheme. The model's governing equations are the Euler Equations. The hydrostatic pressure (Laprise, 1992) is used as the vertical coordinate. The Euler equations are essential to fully benefit from the versatility of the variable mesh since the hydrostatic assumption breaks down at the smaller mesoscales which will be resolvable. For the resolutions foreseen for operational use in the next few years however, the hydrostatic approximation is still a valide one, and the model will optionally make use of it.

If the coding and modular testing of this model is well underway, it is still to soon to be able to show results obtained from the baroclinic version of the model. The results that will follow where obtained with the shallow-water prototype of the model.

2. Results so far

The shallow-water equations support both Rossby and gravity modes, and are typically used to construct prototypes allowing the proof of concept of some of the features of new models.

Our first step was to test the viability of the global variable-mesh strategy described earlier. This was done extensively with the shallow-water prototype.

2.1. VARIABLE RESOLUTION FORECAST

It was found that it is possible to almost exactly reproduce a 2-day forecast on a 80° of longitude x 60° of latitude area using a variable-resolution mesh, at one seventh the cost of running the model with uniformly-high (0.5°) resolution everywhere. Both models had a resolution of 0.5° on the sub-domain (covering North America). Figure 3 shows the RMS height differences between a model run using the uniform high-resolution, and one with the same high-resolution, but only on the region of interest. The differences are very small in the first 48 hours, but increase significantly afterwards. This is not surprising given the low resolution outside the subdomain. After five days, the errors are predictably important, but a study of the geopotential height field (not shown) of the forecast obtained with the variable grid, shows no sign of noise, or shocks, whose presence could be linked to the use of a variable mesh.

The overhead associated with using a model of global extent for short-range forecasting is relatively small: fully 70% of the points in each direction are on the 0.5° uniform-resolution area of interest, and the overhead of using variable resolution outside the area of interest is consequently comparable to that of the sponge regions of one-way interacting models.

The results obtained are very encouraging, and lead us to believe that the baroclinic version of our model will perform well.

The model's variable mesh offers an impressive versatility, allowing the model to be used to perform not only forecasts, but to study the atmosphere in a much broader sense. The reconfigurable grid can be used to study phenomena ranging from global to meso-γ. Because the model has a spherical geometry, problems due to the imposition of lateral boundary conditions and their specification are avoided.

2.2. α AND β MESO-SCALE EXPERIMENTS

Experiments that can be performed at the mesoscale with a shallow-water model are limited, but here are some examples that mainly serve to illustrate the versatility and robustness of the global variable mesh approach.

2.2.1. *geostrophic perturbation.* A mid-latitude area of interest of 1100 x 1100 km^2 is defined, where the resolution is approximately 2.5km. Outside this area, the meshlength increases by a factor of 10%. The result is a grid with 583x567 points of which 60% are concentrated on the area of interest. To the height field taken from a 500 mb analysis, a geostrophically balanced perturbation is added. As can be seen in Fig. 4, the feature is advected with the flow, and leaves the domain with no apparent problem. A study of time-series (not shown) in various points of the height field reveals no problems.

A similar experiment was also done with a much finer mesh: a region of approximately 110x110 km^2 with a resolution of 250 meters was defined. The global grid has 631x617 points, of which 51% are on the small region of interest. The results of this experiment where also very positive: in 2 hours, the geostrophically-balanced perturbation is advected through the domain and, it exits with no apparent shock due to the change in resolution.

2.2.2. *gravitational perturbation.* An experiment with a gravitational perturbation was also done: in an atmosphere at rest, the propagation of a gravitational mode is examined. As can be seen in Fig. 5, the perturbation propates smoothly across the domain, and exits the region of fine resolution with no reflexion or any other apparent problem.

It should be emphasized that these results where obtained with a model that has no explicit dissipation, or diffusion.

2.3. EMERGENCY RESPONSE MODEL

The Canadian Meteorological Centre has a third model: CANERM - the Canadian Emergency Response Model. This model is coupled with one of the existing models to obtain forecasts for pollutant or volcanic ash dispersion, and radioactive clouds advection. CANERM's possibilities are limited by the resolvable dynamical scales of the conveying model. Coupled with the new global model it will have basically limitless possibilities.

2.4. CLIMATE EXPERIMENTS

There is a growing interest in the climate modeling community for experiments on regional climate. The new model, with the appropriate parameterization can be used as a global GCM, and in variable resolution mode, as a regional climate model. The variable mesh enables a smooth embedding of the regional area of interest in the global domain, illiminating the need for sponge layers, and interpolations at the boundaries of the regional area. In the latter case, the major difficulty resides in having a parameterization that is resolution dependent. The way the model is presently constructed for its operational use, two distinct sets of parameterizations are used for its regional and global mode. The regional parameterization is tuned for the area of interest, and therefore is not very good in the outer area, an error recognized, but found to be negligeable for the short term forecasts produced, since it is argued that the error is made far from the area of interest, and therefore will not have time to propagate and contaminate the forecast.

Of course in the context of a long term integration -- and here long term could mean as little as five days -- this sort of error in parameterization is unacceptable. If a resolution dependent parameterization is used however, the global variable mesh model could be an interesting climate model.

There is a project underway with the shallow water prototype of the new model. Its purpose is to examine to what extent is it possible to reproduce for a region the climate obtained by a uniform resolution climate model, using a regional model driven only by the large scale dynamics of a low-resolution climate model. The project addresses the basic assumption which is made in all regional studies that all you need to obtain a good regional climate simulation is a model to provide the large-scale dynamics, and another one to give you the regional surface processes.

The simplified shallow-water context allows for a very high resolution control, and a *perfect* climate model to drive the nested mesoscale model. The nesting of the mesoscale area is done in a continuously varying environment. The importance of the mesoscale features lost by the lack of resolution of the driving model can be mesured by comparing the resulting integrations with those obtained from the uniformely high resolution control. The dependance of the mesoscale response to the resolution of the driving model can also be examined. The results of this study should be published soon.

3. Future work

The baroclinic version of the new global model is now well underway. The dry hydrostatic version will hopefully be ready by the end of the summer, and soon after, the branching of the physical parameterization will begin. Much of the code for the non-hydrostatic version will be done at the same time as that of the hydrostatic version, but its tuning will begin only once the hydrostatic model gives satisfactory results.

4. Conclusions

After extensively testing its shallow-water prototype, the construction of a new non-hydrostatic semi-implicit semi-Lagrangian variable-resolution global model has begun. If this model holds its promises -- and the shallow-water prototype leads us to believe it will -- it will replace the two operational models of CMC, i.e. the spectral and the FER models. It will be a versatile and powerful tool to study meso-scale phenomena, pollutant transport, and even regional climate.

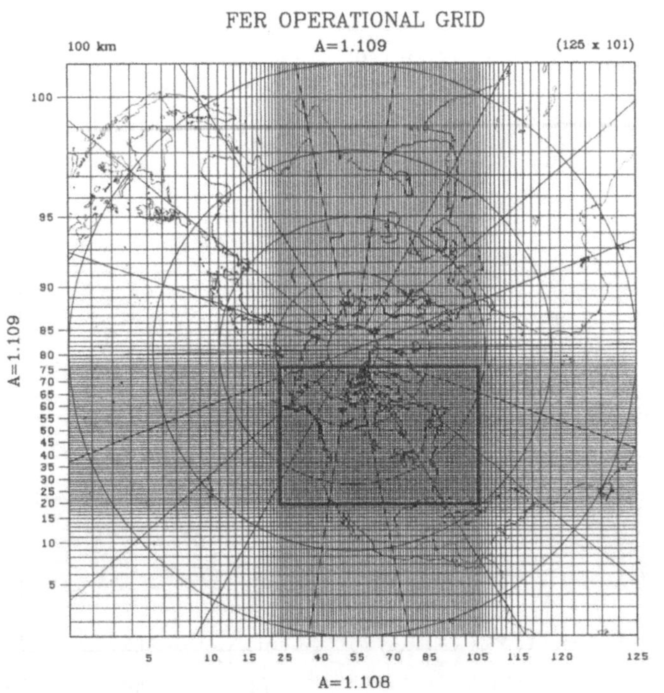

Fig. 1. The 125 x 101 variable-resolution horizontal grid of the Finite Element Regional model on a polar-stereographic projection. Resolution is 100km in the central window (heavy rectangle).

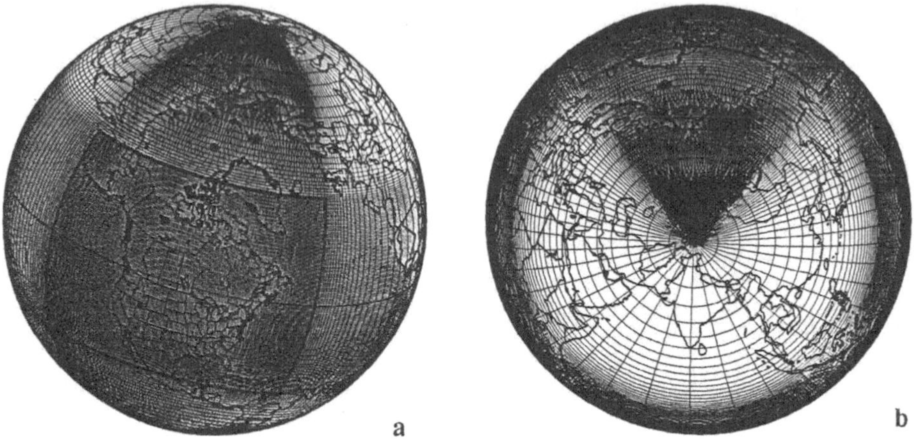

Fig. 2. (a) A variable-resolution 224x169 mesh having an 80° x 60° window of uniform 0.5° resolution, centered on (103°W, 51°N). (b) As in (a), but viewed from a different perspective.

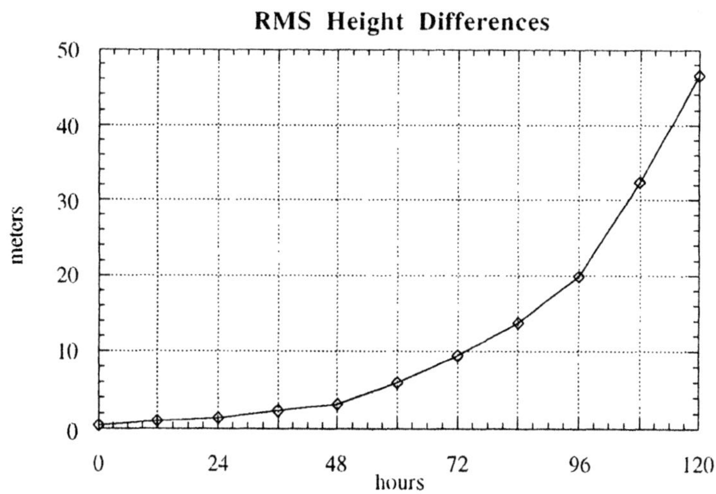

Fig. 3. RMS height differences (m) over the uniform-resolution window between integrations using a varible mesh, and a uniform mesh, as a function of forecast duration (hours).

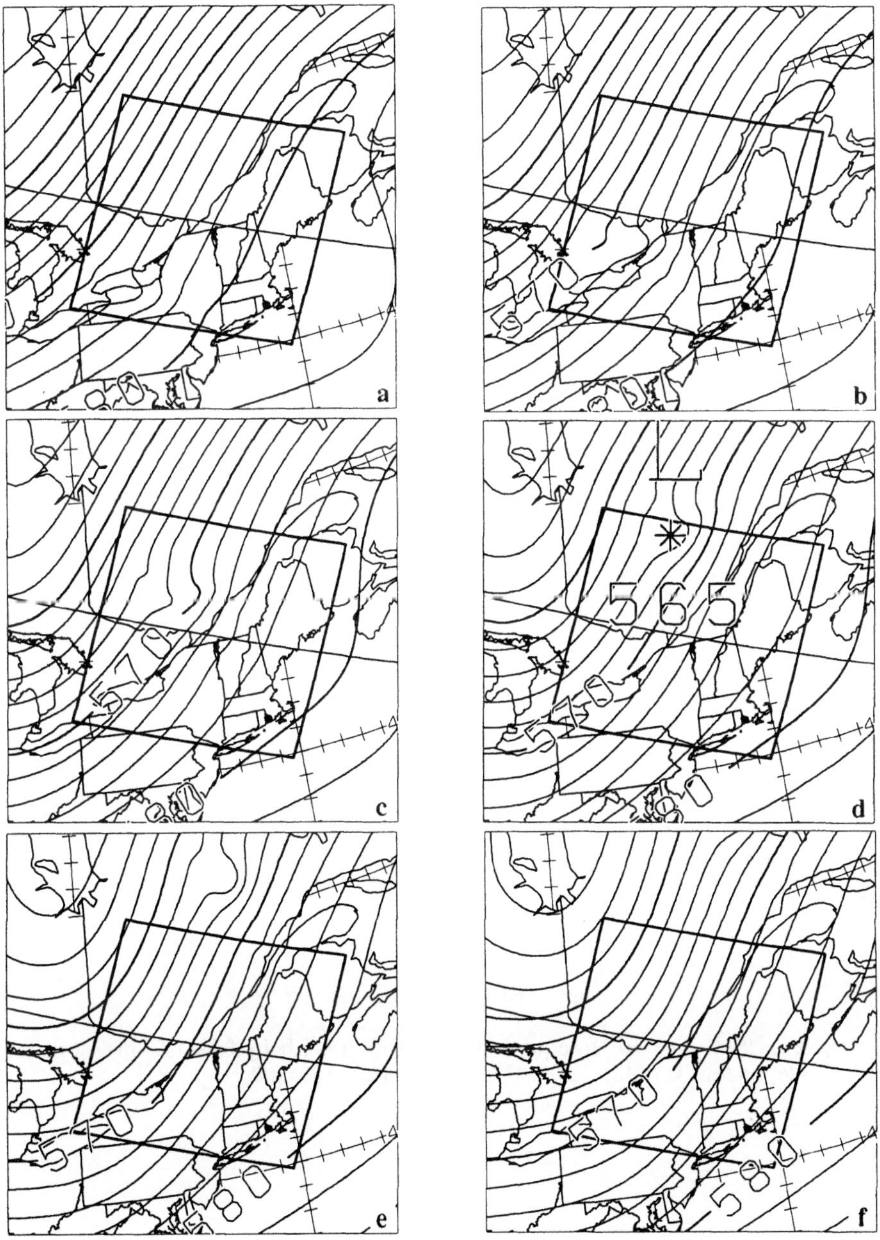

Fig. 4. Time evolution of a geostrophic perturbation at (a) 0 h, (b) 4 h, (c) 8 h, (d) 12 h, (e) 16 h, (f) 20 h.

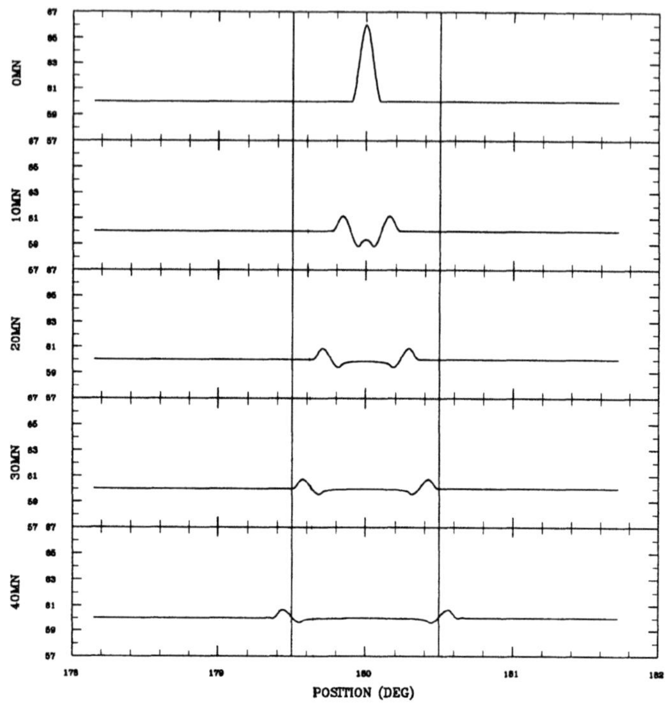

Fig. 5. Domain cross section of the geopotential height field with a gravity wave perturbation. The solid lines mark the limits of the uniform-resolution subdomain of the global variable mesh.

REFERENCES

Laprise, R., 1992: The Euler equations of motion with hydrostatic pressure as an independent variable. *Mon. Wea. Rev.*, **120**, 197-207.

EXPERIENCES WITH PARALLEL PROGRAMMING IN METEOROLOGY

Fritz G. Wollenweber

European Centre for Medium Range Weather Forecasts
Shinfield Park
Reading, RG2 9AX
United Kingdom

email: fwollenweber@ecmwf.co.uk

ACKNOWLEDGEMENTS

The efforts described in this paper are the work of a number of people. First of all Adrian Simmons, Geerd Hoffmann and David Dent need to be mentioned, who are managing the MPP efforts at ECMWF. Then there are the members of the parallel group at ECMWF - present and past - Lars Isaksen, Tuomo Kauranne, Saulo Barros and Guy Robinson. Special appreciation is due to our German collaborators Ute Gaertel, Toni Schueller and Wolfgang Joppisch from GMD, St Augustin, Germany, who did the major part of actually programming the parallel version of IFS available at that time. Last but not least funding by the European Commission (Project Nr. 6643 - PPPE) has to be acknowledged.

INTRODUCTION

The European Centre for Medium-Range Weather Forecasts (ECMWF) is well known as a high performance computing site with considerable computer power installed. Since its beginnings in 1974 ECMWF has been involved in parallel computing because of an early exposure to parallel processing through the Cray line of parallel processor machines.

F.-X. Le Dimet (ed.), High Performance Computing in the Geosciences, 33–41.

Since 1984 ECMWF is organizing a bi-annual workshop on the use of parallel processing in meteorology, with an ever increasing attendance (the next workshop will be held in 1994). A way pursued to build up expertise in parallel processing is the continued participation in EC funded R&D projects, as e.g. GENESIS (1989-91), PPPE (1992-95) and GPMIMD2 (1993-95/96). In 1992 ECMWF initiated the European RAPS benchmarking effort, which so far turned out to be very successful. Finally there are our collaboration efforts with experienced computer science partners like GMD in Germany and CERFACS in France. The efforts at ECMWF are aimed at all aspects of parallel programming, which are of concern for an operational high performance computing centre like ECMWF (e.g. can important codes be ported easily, is a convenient multi-user service possible, is there enough support for the common user to make use of the parallel computer?). This paper will briefly introduce the overall effort into parallel programming and then focus especially on the ongoing PPPE work.

1. GENESIS

The GENESIS project enabled ECMWF to investigate the feasibility of performing weather forecasts with a global spectral weather prediction model on a massively parallel computer. The main goal was to define a good strategy for parallelization, both algorithmically and programming-wise. To achieve that the computational structure and the data access patterns of global spectral weather models have been studied . Different parallel approaches have been considered (parallel pipelines, 2D-decomposition, 3D-transposition) and the transposition approach came up as the method with the most advantages (Barros and Kauranne, 1990). In a next step performance estimates have been calculated under simplified assumptions but considering the cost of communication through startup latency and communication bandwidth. These performance estimates gave us some idea about the performance we are likely to achieve on different computers. To test the reliability of these estimates, simplified versions of the model have been implemented and benchmarked on different platforms (Barros and Kauranne, 1993). The conclusion we can draw from these tests is, that global spectral weather prediction models will run well on massively parallel computers, with good speed-ups and good performances as long as it is possible to achieve good single node performances.

2. GPMIMD2

The EC funded GPMIMD2 project has several aspects. It tries to establish a European presence in the High Performance Computer market. It shall demonstrate a large European MPP in a production environment tackle Grand Challenges problem, like weather prediction and climate modelling. To achieve these goals it will be necessary to port existing physics and industrial applications to distributed memory machines. A further spin-off of this project will be the development of parallel scientific libraries and user packages to be used with MPP computers. For ECMWF the most significant achievement will be the gain of experience in operating a large MPP in a production like environment for considerable time doing real work, like e.g. ensemble forecasting and exploring strategies for computer centres to provide balanced HPC services. Table 1 summarizes the participants and some of the application codes employed.

- MEIKO	
- Telmat Informatique	
- Parsys Ltd	
- CERFACS	
- CERN	High Energy Physics Appl.
- ECMWF/DMN	IFS (weather prediction)
- DMN	satellite applications
- DKRZ	HOPE (ocean model)
- DWD	DEUTSCHLAND model
- UKMO	UNIFIED FCST

Table 1: List of GPMIMD2 participants and application codes.

3. RAPS - Real Applications on Parallel Systems

The RAPS benchmarking effort was initiated by ECMWF in 1992. The idea is to ease the efforts necessary for vendors and potential buyers to benchmark the equipment by providing a suite of applications which are representative for a large class of application programs. Benchmarking all or parts of this suite should then allow to make decisions about what hardware would be best for particular problem areas and code types. The methodology is as follows. The software authors develop portable parallel versions of their production codes, using FORTRAN 77 and PARMACS 5.1 as a programming model. The parallel codes are provided to a working group of hardware vendors for benchmarking. The hardware vendors concentrate on specific optimization. A sound benchmark methodology has to be applied as described by Hockney (1992). RAPS is split into the "RAPS Consortium" which represents the benchmark providers and the "RAPS Consultative Forum", which represents the vendor side. Table 2 lists the members both of the Consortium and of the Consultative Forum.

AVL, Graz	A.C.R.I
CERFACS, Toulouse	CONVEX
CERN, Geneva	Cray Research
Deutscher Wetterdienst, Offenbach	Fujitsu
DKRZ, Hamburg	IBM
ESI, Rungis	Intel
GMD, St. Augustin	Meiko
Meteorologie Nationale, Toulouse	
Meteorological Office, Bracknell	
NERC, Birkenhead	
PALLAS, Bruehl	
University of Southampton	

Table 2: List of members of RAPS Consortium (left) and of RAPS Consultative Forum (right).

PALLAS GmbH in Germany is coordinating the effort. So far benchmark versions of a number of codes have been provided, tutorials about the codes have been held and the codes are currently being implemented and benchmarked on different systems. It is anticipated that towards the end of this year the Consultative Forum will present benchmark results of the different codes for a number of hardware platforms.

4. Lessons learned from MPP work so far

One of the main conclusions that can be drawn so far is that it is far more difficult to program MPP's than it is to program shared memory supercomputers. The main problem is that the data structures and algorithm's employed have to be understood to a far greater extend compared with conventional supercomputers. The adherence to standards

The requirements to be met by an integrated portable development environment are numerous (Willers, 1990). Parallel programming environments have to reduce or even eliminate the effort of reworking application codes to move them from a serial computer to a parallel machine. The risk of introducing new errors into the application has to be avoided. The parallelization must run on the target platform in a efficient and secure manner. The environment should relax dependencies of performance on the optimal choice of data distribution, communication and machine characteristics. It should provide different levels of distributed memory programming as e.g. message passing to exploit coarse grain parallelism and HPF to fine-tune or to exploit fine-grain parallelism. As PPPE aims at the scientific/engineering user community it is essential to support FORTRAN/F90 programming languages. Another area that has to be supported is the problem to control the optimality of the parallel application. For this purpose performance analysis tools have to be provided. Finally it is necessary to be able to integrate into existing environments.

Table 3 lists the participants in PPPE and the application codes used to evaluate the usefulness and quality of the produced toolset. How important adequate programming environments are has been stressed - especially from the users side (Willers,1990) - in the past. The development of several vendor specific environments and also similar projects elsewhere (Gao et al., 1992) underlines the prime role these programming environments will play in the move towards the operational use of massively parallel computers at production sites.

- MEIKO	
- Intel	
- IBM	
- Pallas GmbH	
- GMD	
- University of Southampton	GENESIS/RAPS benchmark suite
- University of Vienna	
- Emeraude	
- NA Software	
- AVL	FIRE combustion engine simulation

Table 3: PPPE participants and application codes

- ECMWF	IFS weather prediction code
- ESI	PAMCRASH crash simulation
- FIRST	ELLPACK PDE solving tool
- Simulog	MS3D/NSTC3D mesh generation tool

Table 3 (continued): PPPE participants and application codes

The runtime system of PPPE will comprise an HPF mapper, which translates HPF code into F77/F90 plus message passing calls, a HPF debugger, a parallel debugger, a F77/F90 compiler and an instrumented message passing system (PARMACS/MPI).

At the beginning of the project a number of tools were already available. They originate partly from other EC-projects, e.g. PCTE, PATOOLS, VFCS, and in part from other sources like the public domain (ParaGraph and PATOP). The tools are listed in the following:

- PCTE
- PARMACS 5.1
- PATOOLS
- PATOP
- ParaGraph
- VFCS

PCTE, which stands for Portable Common Tool Environment serves as the integration platform of PPPE. PCTE is an international standard for tool integration, it is based on a number of new software engineering approaches. PARMACS 5.1 are the GMD/ANL message passing macros which will build the basis for communication on the distributed memory machines envisioned. During the course of the project PARMACS 5.1 will be superseded by PARMACS 6.0, a library based approach for message passing, and ultimately replaced by MPI (Message Passing Interface) the upcoming standard for message passing. PATOOLS, PATOP and ParaGraph are all so called performance analysis tools. They enable the programmer to look at the code from a different point of view and to examine for example the communication pattern and other performance related program features. VFCS stands for the Vienna Fortran Compilation System, which is a automatic parallelization tool. VFCS helps to convert FORTRAN 77 programs into parallel programs using F77 and message passing. It builds mainly on the data parallel paradigm.

5. Examples

In an other talk at this conference (U. Gaertel, 1993) a detailed presentation of our parallel shallow water version of IFS will be given and the reader is referred to this paper to find out more about the performance of IFS on different parallel hardware platforms. In this paragraph some examples on the use of the currently available performance analysis tools will be given. Unfortunately these tools really require colour displays so that it will not be possible to fully demonstrate their capabilities in this paper. The ParaGraph toolset which will form the basis of the performance analysis tools, is a tool originally developed for the public domain. It will be further enhanced and put on a more standardized graphical basis (Motif). The Feynman diagram of this tool is used to demonstrate some of the capabilities which are available. The Feynman diagram is a time -process diagram with the host shown as node 0. Coloured time axes show the current activities of the different nodes (not shown here). The communication between the nodes shows as connecting lines between them, starting at the sending node and terminating at the receiving node (considering increasing time along the axis). Figure 1 shows the communication pattern for a run with the full IFS model on 4 nodes. You see different stages (bursts) of communication and some computation inbetween. There are two big gaps in this time sequence. They represent time-steps where the full radiation model is exercised. The smaller gaps are time steps where only a limited version of the radiation model is used.

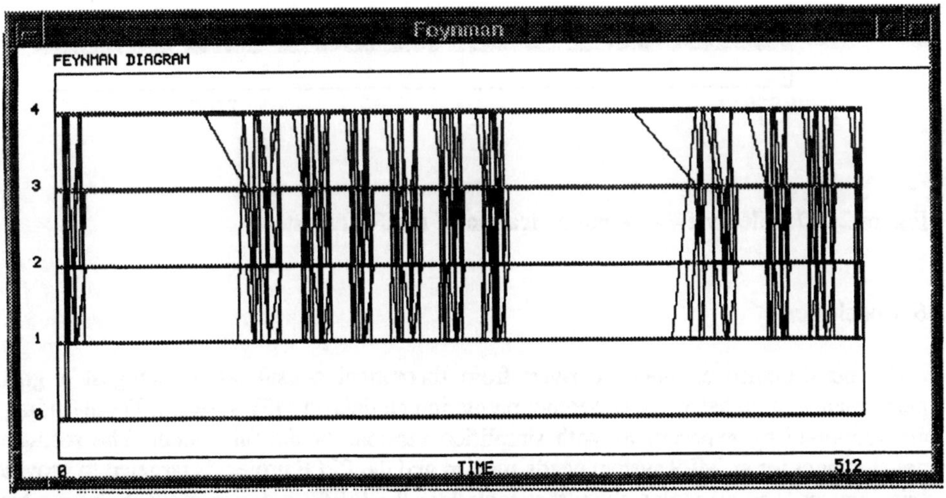

Figure 1: Communication pattern of full IFS model

Figure 2 takes a closer look at the communication between nodes. It is possible to determine the communication characteristics of our code from it. In each time-step all the data is communicated. Every node is communicating with every other node. The communication is organised in a stepwise manner. In a first step a node sends data to the node below it (in the figure) and receives data from the node directly above it. In the second step the node sends data to the node below the one it send to in the first step (2 below) and receives data from the node directly above the node it received from in the first step (2 above). This continues until all the data is communicated.

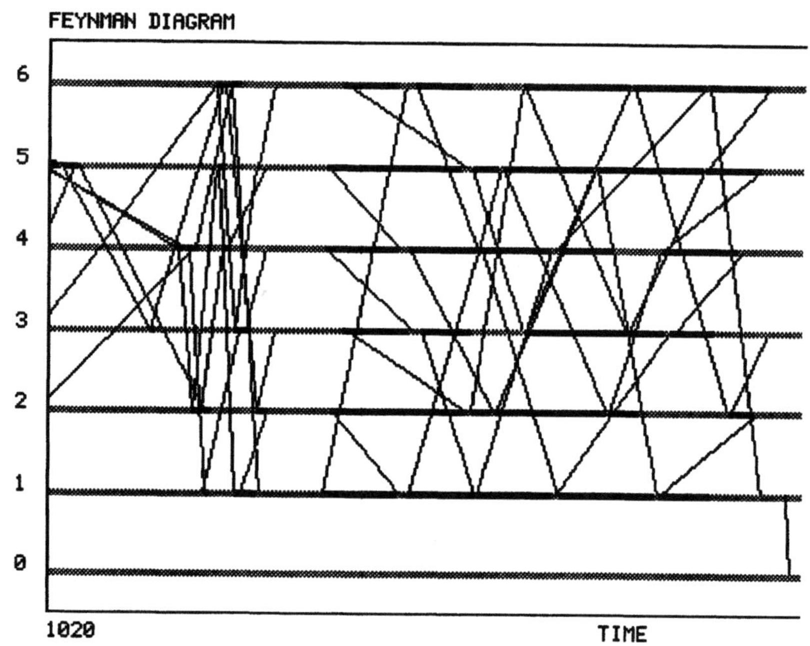

Figure 2: Detailed view of communication of an IFS timestep

6. Conclusion

The performance estimates derived from theoretical considerations suggest a good performance of global spectral weather prediction models on MPP systems. These findings are supported by experiments with simplified versions of the full model. The software development for an MPP system needs support and the PPPE project is targeted to provide this support. The examples show that tools like the PPPE tools are really helpful when dealing with large application codes like IFS.

7. References

Barros, S. R. M. and Kauranne, T.
Spectral and multigrid spherical Helmholtz equation solvers on distributed memory parallel computers
in Proceedings Fourth workshop on "Use of parallel processors in meteorology", 26-30 November 1990, ECMWF workshop proceedings, ECMWF, Shinfield Park, Reading RG2 9AX, UK.

Barros, S. R. M. and Kauranne, T.
On the parallelization of global spectral Eulerian shallow-water models
in Proceedings of the Fifth ECMWF workshop on the Use of parallel processors in meteorology, 23-27 November 1992, Reading, UK
G-R Hoffmann and T Kauranne (Eds.), 1993, World Scientific Publishers, 544 pp

Gaertel, U.
Portable Parallelization of the ECMWF Weather Production Code
in Proceedings "High Performance Computing in the Geosciences", Les Houches, 21-25 June 1993

Gao, G., Hendren, L., Panangaden, P., Feeley, M., Tao, L., Hancu, M., Hum, H., Lebensold, J., and Van Dongen, V.
Towards a Portable Parallel Programming Environment
Proceedings of Supercomputing '92, Montreal, 7-10 June 1992

Hockney, R.
A Framework for Benchmark Performance Analysis
1992, Electronics and Computer Science Department - Tech. Report
Southampton University SO9 5NH, England UK

Willers, I.
State-of-the-art programming environments on parallel computers
in Proceedings Fourth workshop on "Use of parallel processors in meteorology", 26-30 November 1990, ECMWF workshop proceedings, ECMWF, Shinfield Park, Reading RG2 9AX, UK.

Porting the Quasi-Nonhydrostatic Meteorological Model to the Kendall Square Research KSR1

C. F. BAILLIE, A. E. MACDONALD and S. SUN
NOAA Forecast Systems Laboratory
Boulder
CO 80303
USA

ABSTRACT. The Forecast Systems Laboratory is currently developing and testing weather models on various parallel computers with a view toward ultimate production use. As part of that effort we have ported a "Quasi-Nonhydrostatic" Meteorological Model to the Kendall Square Research KSR1 parallel supercomputer. The porting was done in three stages: first the original sequential code was made efficient on one processor, then the code was parallelized using "tiling" directives, and lastly the parallel code was optimized. We describe this process in some detail and report the resulting performances.

1. Introduction

The "Quasi-Nonhydrostatic" (QNH) model is a non-hydrostatic, compressible meteorological model whose dynamics is based on the "approximate system" of Browning and Kreiss [1] plus the incorporation of topography as described in Browning and MacDonald [2]. QNH is a "full physics" model with packages for moist physics, radiation and turbulence. The dynamics part of QNH, known simply as the "well-posed" topographical model (WPT), was ported to the KSR1 by us previously [3]. We have built on the lessons learned there in doing this port of the complete QNH model. In addition, over the last year, we have substantially improved our understanding of the KSR1 architecture and its influence on performance. Therefore we shall begin with a detailed description of the KSR1 hardware as background so we can explain how one optimizes sequential codes on one processor. Next we shall briefly outline the QNH model and code. Then the "tiling" method of parallelization will be described. Finally we give some details on how the resulting parallel code is optimized. Throughout we shall illustrate these steps with performance results for QNH.

F.-X. Le Dimet (ed.), High Performance Computing in the Geosciences, 43–58.

2. The KSR1 and single processor code optimization

The Kendall Square Research KSR1 MPP (massively parallel processor) is the lastest shared-memory multiple-instruction, multiple-data (MIMD) computer [4]. It is built as a hierarchical machine, with a "fat-tree" type architecture, consisting of up to 34 rings of 32 processors. The memory is logically shared as there is one global address space, however it is physically distributed among the processors, each one having 32 MBytes. Most importantly the whole memory is treated as a cache, that is, whenever a processor gets data from another processor's memory or cache, a copy of this data is made in its own cache thereby significantly speeding up subsequent accesses to it. The rings are essentially hardware "search engines" resolving references to addresses that are not found in the processor's local cache by fetching the data automatically – hence all data movement is transparent to the user.

Unfortunately this wonderful feature for ease of programmability leads to some difficulties in obtaining high performance. The main problem is the actual hardware implementation of the cache. As is common in all mircoprocessors today, the KSR1 custom processor chip contains a cache, called the "subcache" in order to distinguish it from the 32 MByte local cache. This subcache is physically 2-way associative. This means that addresses separated by certain "magic numbers" get mapped into the same area of the subcache which can hold at most only two of them. Thus if several of these addresses are accessed by the program one after another, it gives rise to the problem known as "subcache thrashing" which seriously degrades performance. This pattern of address reference is in fact exactly what occurs in a Fortran program stepping through the second (or higher) dimension of an array (since Fortran arrays are stored in column-major order). The trivial fix for this problem is, of course, not to have arrays whose first dimension size leads to a magic number. This is most easily done by picking odd numbers which are not a power of two. Therefore in porting a code to the KSR1 the very first step is to change array declarations like $a(128, 128, 32)$ to $a(129, 129, 32)$ or equivalently $a(0 : 128, 0 : 128, 32)$. Coincidentally this is precisely what is done in codes designed for vector computers like the Cray but for a different reason, namely in order to avoid memory bank conflicts.

There is another single processor optimization which is crucial for certain codes like QNH on cache-based machines. This comes about because in caches, what is cached is not an individual variable of the program but a "cache line" which contains several variables. Therefore optimum use of the cache is obtained when the program accesses all of the variables on a cache line one after another before moving to the next cache line. This necessitates re-structuring conventionally written sequential scientific Fortran programs which typically update one variable or array element, then move onto another, then another, etc, as in the following example (taken from QNH) :

```
do k = 1,nz
 do j = 1,ny
  do i = 1,nx
   ffu = ffu + mu * (u(i+1,j,k) + u(i-1,j,k)
       + u(i,j+1,k) + u(i,j-1,k) - 4*u(i,j,k))
   ffv = ffv + mu * (v(i+1,j,k) + v(i-1,j,k)
       + v(i,j+1,k) + v(i,j-1,k) - 4*v(i,j,k))
   ffw = ffw + mu * (w(i+1,j,k) + w(i-1,j,k)
       + w(i,j+1,k) + w(i,j-1,k) - 4*w(i,j,k))
       + muz * (w(i,j,k+1) + w(i,j,k-1) - 2*w(i,j,k))
  end do
 end do
end do
```

We see that first a cluster of elements of u is modified, then similar clusters for v and w are modified. However, $v(i,j,k)$ and $w(i,j,k)$ are stored a long way from $u(i,j,k)$ in memory so there is not much data locality for the cache to take advantage of.

Instead the array elements should be "aggregated" or "coallesced" by declaring another array, say, $Q(3,0:nx,0:ny,nz)$ and copying $u(0:nx,0:ny,nz)$ into $Q(1,0:nx,0:ny,nz)$, $v(...)$ into $Q(2,...)$ and $w(...)$ into $Q(3,...)$. For readability one can also declare parameters $U=1, V=2, W=3$ so that $u(...)$ in the original code is now referred to as $Q(U,...)$. In order to do this aggregation simply without making mistakes some kind of program transformation tool is very helpful. We have been using the Sage++ tool from Gannon's group at Indiana [5]. To this we added the "user annotation" 'C$ann[Aggregate(u,v,w)]' which accomplishes the transformation just described, producing the following code :

```
do k = 1,nz
 do j = 1,ny
  do i = 1,nx
```

```
        ffu = ffu + mu * (Q(U,i+1,j,k) + Q(U,i-1,j,k)
            + Q(U,i,j+1,k) + Q(U,i,j-1,k) - 4*Q(U,i,j,k))
        ffv = ffv + mu * (Q(V,i+1,j,k) + Q(V,i-1,j,k)
            + Q(V,i,j+1,k) + Q(V,i,j-1,k) - 4*Q(V,i,j,k))
        ffw = ffw + mu * (Q(W,i+1,j,k) + Q(W,i-1,j,k)
            + Q(W,i,j+1,k) + Q(W,i,j-1,k) - 4*Q(W,i,j,k))
            + muz * (Q (W,i,j,k+1) + Q(W,i,j,k-1) - 2*Q(W,i,j,k))
      end do
    end do
  end do
```

Now when the cluster of addresses around $Q(U, i, j, k)$ is fetched, similar addresses for $Q(V, i, j, k)$ and $Q(W, i, j, k)$ are pulled in on the same cache line (which on the KSR1 contains 16 64-bit variables) ready to be used next. In our codes this rearrangement of data yields up to a factor of two improvement in performance on the KSR1. We shall return to the actual performance numbers after outlining the QNH model and code.

3. The QNH Model and Code

The Quasi-Nonhydrostatic (QNH) meteorological model, like all regional forecast models, consists of two main parts: dynamics and physics. We already discussed the dynamics part of QNH in [3] so will not repeat it here; instead we just list the basic equations:

$$\frac{du}{dt} = -\frac{1}{\rho_o}\frac{\partial p}{\partial x} + fv + \nu\nabla^2 u \tag{1}$$

$$\frac{dv}{dt} = -\frac{1}{\rho_o}\frac{\partial p}{\partial y} - fu + \nu\nabla^2 v \tag{2}$$

$$\frac{dw}{dt} = \alpha(-\frac{1}{\rho_o}\frac{\partial p}{\partial z} + g\theta - \frac{gp}{\gamma P_o}) + \nu\nabla^2 w \tag{3}$$

$$\frac{dp}{dt} = -\gamma P_o(\frac{\partial u}{\partial x} + \frac{\partial v}{\partial y} + \frac{\partial w}{\partial z}) + g\rho_o w + \nu\nabla^2 p \tag{4}$$

$$\frac{d\theta}{dt} = -\frac{1}{\theta}\frac{\partial \theta}{\partial z}w + \nu\nabla^2\theta \tag{5}$$

where

$$\alpha = (\frac{\Delta z}{\Delta x})^2 \tag{6}$$

is the "speed reduction" constant, which mathematically makes the equations symmetric and physically slows vertical sound waves down, so that the speed of sound in vertical and horizontal grid space is similar. u, v, w are the winds in the three dimensions, p is pressure, θ is temperature, ρ_0 is mean density in the vertical, ν is horizontal diffusion coefficient, f is Coriolis parameter, g is gravity constant, and γ is adiabatic exponent.

Recently, however, the dynamics part was modified to include well-posed boundaries which make the code somewhat harder to parallelize so we shall describe this new section of the dynamics. The theory of boundary conditions for the open boundary problem for symmetric hyperbolic systems was developed by Browning and Kreiss [6]. The so-called well-posed boundary conditions are based on the characteristics of the partial differential equations on each boundary. Consider a simple hyperbolic system describing sound wave propagation in a fluid moving at speed U_0. Let the speed of sound, $c >> U_0 > 0$. On the western boundary, two characteristics, $U_0 + c$ going from west to east, and $U_0 - c$ going from east from west, bring in the information from the exterior (either an outer coarse-mesh model or data) and from the interior, respectively. The combination of these two characteristics yields the final boundary values of u and p. If superscript c represents values obtained from the coarse mesh or data, n values from the inner numerical model, and $*$ the final boundary values, we have

$$u^* + p^*/c = u^c + p^c/c \quad ; \quad u^* - p^*/c = u^n - p^n/c \tag{7}$$

Thus the well-posed boundary values u^* and p^* are obtained by solving these equations.

The physics package in QNH currently includes moist physics, radiation and turbulence. The moist physics package is based on Schultz [7]. It contains six "moist variables": water vapor, cloud liquid, cloud ice, rain, snow and precipitated ice. The limited radiation package currently calculates heating associated with radiation. The Mellor-Yamada [8] Level 2.0 turbulence parameterization is used here for the turbulence in the boundary layer of a few kilometers thick.

The structure of the QNH code is as follows:

```
Loop over time

 Loop over space to calculate forcing functions
   from dynamics including advection, topography,
   horizontal diffusion, and physics
```

```
    Loop over space for dynamics time differencing
    Loops over horizontal boundaries to make them well-posed
    Loop over top vertical boundary to make it well-posed
    Loop over space for moist physics advection
    Loop over space for moist physics time differencing
    Loop over space for moist physics phase change
    Loop over space for radiation
    Loop over space for turbulence

  End loop over time
```

All of the loops over space go over the whole horizontal plane $nx \times ny$ but they have different extents in the vertical. For dynamics the vertical loop goes all the way to the top of the atmosphere nz; for moist physics and radiation it goes approximately half-way; and for turbulence it is confined to a few layers near the ground. Hence when we come to parallelize the code we shall divide the three-dimensional space up into vertical columns, keeping all the various vertical loops local to each processor.

The original code was developed on a Sun Sparc 2 workstation which does not have any complicated caching so no care was taken in array size declaration (to avoid subcache thrashing) and no array aggregation was done (to improve use of the subcache), as explained above in Section 2. Therefore the original code ran badly on one processor of the KSR1, it also ran badly on the DEC 3000/400 workstation which is based on the DECchip 21064 or "Alpha" chip, as can be seen from the times labeled original in Table I. However, when the code was optimized by adding an extra row to the x and y dimensions of all the arrays and by aggregating the arrays for the primary variables, the speed on the KSR1 and on the DEC workstation increased by almost a factor of 2 – see the last column of Table I. We present the times for the DEC workstation to make the point that todays high performance workstations can benefit from simple code changes which improve data locality in their caches *in exactly the same way* as the KSR1 does. In fact the DEC Alpha chip is what Cray Research has chosen as processor in their first MPP product, the T3D. Thus the programming methodology we are developing for the KSR1 has wider applicability. We would also like to note that the clock speed of the KSR1 is only 20 MHz whereas that of the DEC 3000/400 is 133 MHz. Taking this into account we see that the KSR1 is actually performing better relatively, or putting it another way, speeding up the clock on the KSR1 will significantly improve its single processor performance relative to high end workstations.

TABLE I
Times in seconds for a short run of the original and the optimized versions of QNH with only dynamics and well-posed boundaries using a 32x32x21 grid on three different processors; the KSR1 is for a single processor.

Processor	original	optimized
Sun Sparc 2	31.1	30.4
KSR1	28.6	15.7
DEC 3000/400	7.9	4.5

We instrumented the code (again using the Sage++ tool) in order to discover that 15.7 seconds on the KSR1 corresponds to 7.7 Mflops. Here we have timed only the dynamics and well-posed boundaries to verify that the performance is about the same as what we obtained for the WPT code which is essentially just the dynamics part of QNH. In fact, for WPT [3] after array aggregation we obtained 7.8 Mflops. We actually went further with WPT and rearranged the code to bring updates to the same variables closer together thereby improving the performance to 8.6 Mflops. We could do the same here for QNH, but we elected to investigate parallel performance first.

4. Parallel Implementation on the KSR1

KSR provides several compiler directives in Fortran to aid parallelization of programs, the main one being "tiling" [9]. Essentially, tiling means splitting up a series of nested loops into blocks, each one of which can be executed independently of the others. In order to do this, all the user need do is add a few compiler directives. In our previous work on WPT [3] we explained tiling in some detail and tested various strategies to discover the best for our code. We shall not repeat all that here, suffice it to say that the conclusions for WPT apply to QNH. Namely that it is most efficient to tile the horizontal loops over nx and ny, with the vertical loop over nz in-processor. This horizontal tiling should be done using the so-called "slice" strategy which makes one tile per processor, and these tiles should be rectangular – long in the first dimension and thin in the second dimension (again due to Fortran array storage order). In fact highest performance is obtained when the tiles are $nx \times (ny/N)$, where N is the number of processors.

Of course in order to tile our loops over nx and ny we require a loop nest containing at least these two loops (and possibly also a loop over nz). Therefore we immediately have a problem with the well-posed horizontal boundary loops which go over only nx or ny but not both. The code for the 'Loop over space for dynamics time differencing', the 'Loops over horizontal boundaries to make them well-posed' and the 'Loop over top vertical boundary to make it well-posed' looks as follows :

```
do  k = 1,nz

TILE(i,j)
   do  j = 1,ny
     do  i = 1,nx
       Dynamics time differencing
     enddo
   enddo
   do  j = 1,ny
     Well-posed horizontal i = 1 bdy          ! west
   enddo
   do  j = 1,ny
     Well-posed horizontal i = nx bdy         ! east
   enddo
   do  i = 2,nx-1
     Well-posed horizontal j = 1 bdy          ! south
   enddo
   do  i = 2,nx-1
     Well-posed horizontal j = ny bdy         ! north
   enddo

enddo

TILE(i,j)
   do  j = 2,ny-1
     do  i = 2,nx-1
       Well-posed vertical k = nz bdy         ! top
     enddo
   enddo
```

We have indicated where the KSR tile directives would go with $TILE(i,j)$. Notice that the range of the second tile statement is two less than that of the first (2 to $ny-1$ and 2 to $nx-1$ rather than 1 to ny and 1 to nx). This could lead to unnecessary data movement between processors whose tiles change size unless both loops were included in an $AFFINITY REGION$ statement. This statement tells the compiler to keep the data fixed in the processors at the expense of having different tile sizes. Another way around this is to change the second set of loops to go over the whole $nx \times ny$ plane with an if statement inside to skip the edges; this is what we shall use. The reader may be wondering why we do not simply have one dimensional tile statements for the well-posed horizontal loops (i.e. $TILE(i)$ or $TILE(j)$). The problem with this is that it would cause a great deal of data to be moved around between processors since the two dimensional loops require rectangular tiles whereas the one dimensional loops require strips. Moreover for the i-loop these strips would be in one direction and for the j-loop in the other direction so effectively a data transpose would be happening between the i-loops and j-loops, in addition to the data movement between these loops and the two dimensional loops. The best solution is to obtain two-dimensional loops nested appropriately for tiling by rearranging the code as follows :

```
do  k = 1,nz

TILE(i,j)
   do  j = 1,ny
     do  i = 1,nx
       Dynamics time differencing
     enddo
   enddo

TILE(i,j)
   do  j = 1,ny
     do  i = 1,nx
       if (i .eq. 1) then
         Well-posed horizontal i = 1 bdy             ! west
       endif
       if (i .eq. nx) then
         Well-posed horizontal i = nx bdy            ! east
       endif
       if (i .ne. 1 .and. i .ne. nx) then
         if (j .eq. 1) then
           Well-posed horizontal j = 1 bdy           ! south
         endif
         if (j .eq. ny) then
           Well-posed horizontal j = ny bdy          ! north
         endif
```

```
            if (j .ne. 1 .and. j .ne. ny) then
              if (k .eq. nz) then
                Well-posed vertical k = nz bdy        ! top
              endif
            endif
          endif
        enddo
      enddo

    enddo
```

All we have done is change the one dimensional loops into two dimensional loops plus if statements to select only the one dimensional parts we wanted in the first place. However this minor change is *essential* in order to be able to tile the loops with no data movement and achieve decent parallel performance. This is obvious from the times for the dynamics and well-posed boundaries part of the code listed in Table II. On, for example, 8 processors the two dimensional loop version obtains a speedup of 5.0 whereas the one dimensional loop version manages only 2.6. Notice that in addition to combining all four well-posed horizontal loops we also combined the well-posed vertical loop with them. This saves a tile statement with its associated overhead. We could go further and eliminate one more tile statement by combining our new loop over all the well-posed boundaries with the previous loop over the 'dynamics time differencing'. However this would change the order in which the boundaries are updated and therefore change the results of the code slightly. Nevertheless we tried doing this but as it only sped the code up by a few percent we decided it was not worth it after all. One last thing we tried was getting rid of the if statements to see how much time they were taking up – again it was only on the order of a percent. Incidentally, this rewrite of one-dimensional loops as two-dimensional loops will also help in porting the code to distributed memory message passing MPPs.

Now that we have the code rewritten entirely in the form of two dimensional loops we can go ahead and tile them all the same way to obtain good parallel performance. We show times for all seven loops, plus the total time, in Table III for a small run of the complete (dynamics and physics) QNH code on various numbers of KSR1 processors. In order to better see how efficiently the loops are being parallelized we calculate the speedups from these times, as shown in Table IV. It is clear that despite the overall speedups being fairly good (this is after all a rather small grid so there is not much parallelism left by the time we reach 16 processors) a large fraction of the loops are not yielding satisfactory speedups in themselves. Fortunately these

TABLE II
Times in seconds for a short run of the one dimensional loop (1D) and two dimensional loop (2D) versions of QNH with only dynamics and well-posed boundaries using a 32x32x21 grid on different numbers of processors of the KSR1.

Processors	1D	2D
1	15.6	15.7
2	11.8	8.4
4	7.4	4.4
8	5.0	2.6
16	4.1	1.5

TABLE III
Times in seconds for a short run of the complete QNH code using a 32x32x21 grid on different numbers of processors of the KSR1.

Loop	1 proc	2 procs	4 procs	8 procs	16 procs
Dynamics	12.7960	6.6405	3.4505	1.9175	1.0738
Dynamics time diff	3.1174	1.8070	1.1498	0.7429	0.4272
Moist physics adv	0.2422	0.1572	0.1146	0.0578	0.0466
Moist time diff	0.0589	0.0394	0.0202	0.0134	0.0116
Moist phase change	1.8627	0.9997	0.5149	0.2571	0.1370
Radiation	0.0192	0.0165	0.0099	0.0059	0.0045
Turbulence	0.0374	0.0243	0.0154	0.0073	0.0074
Total	18.15	9.73	5.33	3.06	1.77

loops consume only about 2% of the sequential run time. Most of the time (70%) is spent in the 'Dynamics' loop which shows the second best speedup. Next comes the 'Dynamics time difference' loop (which includes the two-dimensionalized well-posed boundary loop as described above) with 17% of the run time showing third best speedup. And last with 10% is the 'Moist physics phase change' loop which surprisingly shows the best speedup. We say surprisingly because this loop contains a great many if statements to implement all the possible phase changes between the six forms of moisture present. Therefore this part of code has the potential for the most load imbalance since areas of the sky with clouds will require more computing than blue sky. In tests with cloud covering one third of the sky this load imbalance was found to be less than 10% of the time spent in this loop.

Despite the remaining loops using such a small fraction of the run time we will discuss each of them in turn to explain why they show such poor speedups and indicate how to improve them. Firstly, the 'Radiation' loop

TABLE IV
Speedups from times in Table III.

Loop	1 proc	2 procs	4 procs	8 procs	16 procs
Dynamics	1.00	1.93	3.71	6.67	11.92
Dynamics time diff	1.00	1.73	2.71	4.20	7.30
Moist physics adv	1.00	1.54	2.11	4.19	5.20
Moist time diff	1.00	1.49	2.92	4.40	5.08
Moist phase change	1.00	1.86	3.62	7.25	13.60
Radiation	1.00	1.16	1.94	3.25	4.27
Turbulence	1.00	1.54	2.43	5.12	5.05
Total	1.00	1.87	3.41	5.93	10.25

TABLE V
Speedups for a short run of the complete QNH code using a 32x32x41 grid on different numbers of processors of the KSR1.

Loop	1 proc	2 procs	4 procs	8 procs	16 procs
Dynamics	1.00	2.24	4.50	8.45	14.50
Dynamics time diff	1.00	1.86	3.46	5.51	8.48
Moist physics adv	1.00	2.21	4.56	6.26	10.53
Moist time diff	1.00	1.95	3.90	7.62	9.05
Moist phase change	1.00	2.06	4.16	8.24	16.07
Radiation	1.00	1.76	3.75	5.28	7.24
Turbulence	1.00	1.47	3.88	7.30	11.22
Total	1.00	2.07	3.90	6.36	12.54

contains only a couple of statements so the overhead in tiling this loop is a large fraction of its run time. Moreover this overhead increases as the number of processors increases thereby limiting the speedup. The solution to this problem is to combine this loop with other loops in order to reduce the number of tile statements. Similarly the 'Moist physics advection' and 'Moist physics time difference' loops contain only a handful of statements, each one updating a different variable, so not enough work and little data reuse limits their speedup. In addition the vertical extent for these loops is less than half-way to the top of the atmosphere and this is the dimension which is in-processor so again there is not so much work to do for each tile creation and destruction. Lastly, the 'Turbulence' loop has two problems: it is the only loop in the code which has the vertical loop inside the horizontal loops, and this vertical loop runs over only two layers. Having the vertical loop inside the horizontal loops means that large strides are taken in the variables

TABLE VI
Times in seconds for a short run of the complete QNH using various sizes of grids on different numbers of processors of the KSR1.

Processors	32x32x21	64x64x21	128x128x21
1	18.15	100.3	706.9
2	9.73	47.3	319.9
4	5.33	23.4	133.9
8	3.06	12.4	60.2
16	1.77	6.9	29.3
32	-	4.8	18.8
64	-	-	12.1

TABLE VII
Speedups from times in Table VI.

Processors	32x32x21	64x64x21	128x128x21
1	1.00	1.0	1.0
2	1.87	2.1	2.2
4	3.41	4.3	5.3
8	5.93	8.1	11.7
16	10.25	14.5	24.1
32	-	20.9	37.6
64	-	-	58.4

(since the vertical index is the last one) which is bad for the subcache. And again with only two iterations of the vertical loop there is not enough work to ameliorate the overheads. To verify that the speedups would be better if there were more work in the loops we ran a grid of 32x32x41 with four layers for the turbulence. With this Table IV becomes Table V with much better speedups. In fact the two loops which exhibited the best speedups with 21 points in vertical now display superlinear speedups with 41 points. The 'Moist physics advection' loop shows the most dramatic improvement. The 'Radiation' is still the worst but is by far the smallest part of the code. Thus for maximum parallel performance it is important to have as many points in the vertical as possible. This corresponds physically to a high resolution in the vertical which is precisely what models like QNH are aiming for.

Finally we investigate how well the complete QNH code using various sizes of grids scales on different numbers of KSR1 processors. In Tables VI and VII we list the total run times and the speedups, respectively, for 32x32x21, 64x64x21 and 128x128x21 grids on up to 64 processors of the

TABLE VIII
Total Mflops from times in Table VII.

Processors	32x32x21	64x64x21	128x128x21
1	7.0	5.2	3.0
2	13.3	10.9	6.6
4	23.8	22.4	15.9
8	42.7	42.1	35.1
16	70.7	75.4	72.3
32	-	108.7	112.8
64	-	-	175.2

TABLE IX
Mflops per processor from times in Table VIII.

Processors	32x32x21	64x64x21	128x128x21
1	7.0	5.2	3.0
2	6.7	5.5	3.3
4	6.0	5.6	4.0
8	5.3	5.3	4.4
16	4.4	4.7	4.5
32	-	3.4	3.5
64	-	-	2.7

KSR1. We also convert run times to Mflops, the standard unit of performance, in Table VIII. With our instrumented version of the code we find that 18.2 seconds run time on one processor for the 32x32x21 grid corresponds to 7.0 Mflops. This is less than the 7.7 Mflops we obtained above for the dynamics only because the physics has more logical operations. Therefore a 64x64x21 grid, which requires four times as many flops, running in 100.3 seconds on one processor yields 5.2 Mflops. Similarly a 128x128x21 grid taking 706.9 seconds gives 3.0 Mflops. These performances for the larger sizes of grid are poor because the data does not fit in one processor's memory. However since the KSR1 is a shared memory machine the code still runs with the data being swapped to the memory of other processors. When the number of processors is such that the data just fits in their combined memory we obtain a peak in the Mflops per processor, which is shown in Table IX. For the three problem sizes this peak is for 1,4 and 16 processors respectively. We also obtain super-linear speedups (see Table VII) until we reach this number of processors. Unfortunately the Mflops per processor falls off significantly on large numbers of processors. This is true even when

the amount of work per processor is kept constant: comparing 32x32x21 on 1 processor, 64x64x21 on 4 processors and 128x128x21 on 16 processors in Table IX reveals that the Mflops per processor falls from 7.0 to 5.6 to 4.5. We are still not entirely sure why this happens (similar behavior was seen for WPT [3]) – but it is probably due to increased contention in the rings. Of course, the most important (some would say the only important) measure of performance, run time, decreases by about 60 on 64 processors. Therefore the KSR1 is very effective for running weather forecast codes in parallel.

The basic lesson we have learned in parallelizing code for the KSR1 is to minimize the number of tiled loops. Unfortunately this is just the opposite of what one does when vectorizing a code for the Cray. There the goal is to make every loop a vector loop over a few variables. However, as we discussed in Section 2, at least optimizing sequential codes for the two machines is similar.

5. Conclusions and Further Work

We have shown that it is fairly easy to port the QNH model to the KSR1 parallel supercomputer, obtaining reasonable sequential and parallel performance by making minor rearrangements of the data and code. We found a program transformation tool (such as Sage++) very useful in making these rearrangements. Our porting strategy is a three step process: first the original sequential code is made efficient on one processor, then the code is parallelized using "tiling" directives, and lastly the parallel code is optimized. The first step consists of three parts. Firstly an extra row and column is added to all the array dimensions to avoid subcache thrashing. Secondly arrays are aggregated (using solely directives to the program transformation tool) to help with data locality. Thirdly the code is rearranged to bring updates to the same variables closer together (we did not so this here but will do it before beginning production runs of QNH on the KSR1). The second step, namely adding tile statements, turns out to be the most straightforward. The last step is actually the most time consuming one as it involves rewriting one-dimensional boundary loops in two-dimensional form. As we pointed out the sequential optimization part of this strategy renders the sequential code efficient on other cache-based processor chips like the DECchip 21064 and on vector computers like the Cray; and the parallel optimization part helps with parallelization for other MPPs.

Our immediate future plans are to port the QNH model to the Intel Paragon distributed memory MPP. This involves a significant rewrite of the code but fortunately our colleagues have already done this for WPT (see [10] in these proceedings). Moreover they produced the so-called "nearest neighbor tool" which makes it much easier to do this rewrite by hiding a great deal of the message passing. Over the longer term, we are also involved in a project to port the National Meterological Center forecast model, called Eta [11], to MPPs. Based on our experience with WPT and QNH, we have decided to first port Eta to the KSR1 and then to the Intel Paragon, using our three part strategy.

Acknowledgements

CFB was funded in part by the NOAA Forecast Systems Laboratory, and by AFOSR Grant AFOSR-89-0422 with the Center for Applied Parallel Processing (CAPP) at CU.

References

[1] G. L. Browning and H.-O. Kreiss, 'Scaling and computation of smooth atmospheric motions', *Tellus* **38A** (1986) 295-313.
[2] G. L. Browning and A. E. MacDonald, 'Incorporating topography into the multiscale systems for the atmosphere and oceans', *Dynamics of Atmosphere and Oceans* **18** (1993) 119-149.
[3] C. F. Baillie and A. E. Macdonald, 'Porting the well-posed topographical meteorological model to the KSR parallel supercomputer', in: *Proc. Fifth ECMWF Workshop on the Use of Parallel Processors in Meteorology: Parallel Supercomputing in Atmospheric Science*, eds. G.-R. Hoffmann and T. Kauranne (World Scientific, London, 1993).
[4] 'KSR Parallel Programming' manual, *Kendall Square Research, 170 Tracer Lane, Waltham, MA 02154* (October 1991).
[5] D. Gannon, F. Bodin, S. Srinivas, N. Sundaresan, S. Narayana and J. Gotwals, 'Sage++, An Object Oriented Toolkit for Program Transformations', *Technical Report, Dept. of Computer Science, Indiana University* (1993).
[6] G. L. Browning and H.-O. Kreiss, 'Initialization of the shallow water equations with open boundaries by the bounded derivative method', *Tellus* **34** (1982) 334-351; and 'Scaling and computation of smooth atmospheric motions', *Tellus* **38A** (1986) 295-313.
[7] P. Schultz, 'Development and tests of a cloud physics parameterization for real-time aviation and public numerical weather forecasting', *NOAA Technical Memorandum ERL-FSL 6, U.S. GPO: 1993-774-025/89026* (1993).
[8] G. L. Mellor and T. Yamada, 'A hierarchy of turbulence closure models for planetary boundary layers', *J. Atmos. Sci.* **31** (1974) 1791-1806; and 'Development of a turbulence closure model for geophysical problems', *Rev. Geophys.* **20** (1974) 851-875.
[9] 'KSR Fortran Programming' manual, *Kendall Square Research, 170 Tracer Lane, Waltham, MA 02154* (October 1991).
[10] B. Rodrigues, L. Hart and T. Henderson, 'Performance and Portability in Parallel Computing: A Weather Forecast View', these proceedings.
[11] Z. I. Janjic, 'The step-mountain coordinate: Physical package', *Mon. Wea. Rev.* **118** (1990) 1429-1443.

Automatic Differentiation, Tangent Linear Models, and (Pseudo)Adjoints[1]

Christian H. Bischof

Mathematics and Computer Science Division
Argonne National Laboratory
9700 S. Cass Avenue, Argonne, IL 60439-4843
bischof@mcs.anl.gov

Abstract

This paper provides a brief introduction to automatic differentiation and relates it to the tangent linear model and adjoint approaches commonly used in meteorology. After a brief review of the forward and reverse mode of automatic differentiation, the ADIFOR automatic differentiation tool is introduced, and initial results of a sensitivity-enhanced version of the MM5 PSU/NCAR mesoscale weather model are presented. We also present a novel approach to the computation of gradients that uses a reverse mode approach at the time loop level and a forward mode approach at every time step. The resulting "pseudoadjoint" shares the characteristic of an adjoint code that the ratio of gradient to function evaluation does not depend on the number of independent variables. In contrast to a true adjoint approach, however, the nonlinearity of the model plays no role in the complexity of the derivative code.

Keywords: Automatic Differentiation, Adjoint, Tangent Linear Model, MM5, ADIFOR, SparsLinC, Data Assimilation.

1 Introduction

Let $F(p)$ denote the (vector-valued) output of a model F produced by a particular (vector-valued) input p. Employing the Taylor expansion of F around a reference state p_o, we have

$$F(p_o + \Delta p) = F(p_o) + \frac{\partial F(p_o)}{\partial p}\Delta p + \frac{1}{2}(\Delta p)^T \frac{\partial^2 F(p_o)}{\partial p^2}\Delta p + HO(p_o, \Delta p), \tag{1}$$

where the higher-order terms $HO(p_o, \Delta p)$ satisfy $||HO(p_o, \Delta p)|| = O(||\Delta p||^3)$. Hence, the value of the first- and second-order derivatives (we also interchangeably use the terms first- and second-order sensitivities) allows us to derive a first- or second-order approximation of the change of F in response to a perturbation of the input p from its base state p_o. In particular, the first-order Taylor series approximation

$$F(p_o) + \frac{\partial F(p_o)}{\partial p}\Delta p \tag{2}$$

provides a linear approximation to the (usually nonlinear) behavior of F around the point p_o.

[1]This work was supported by the Office of Scientific Computing, U.S. Department of Energy, under Contract W-31-109-Eng-38, by the National Aerospace Agency under Purchase Order L25935D, and by the National Science Foundation, through the Center for Research on Parallel Computation, under Cooperative Agreement No. CCR-9120008.

F.-X. Le Dimet (ed.), High Performance Computing in the Geosciences, 59–80.

Derivatives provide a way for computing a relatively simple approximation of F, thus allowing one to inexpensively explore the behavior of f in the neighborhood of p_o. Hence, derivatives are ubiquitous in numerical computing.

There are four main approaches to computing derivatives:

By Hand: One can differentiate the code by hand and thus arrive at a code that also computes derivatives. However, handcoding of derivatives for a large code is a tedious and error-prone process, although it is likely to result in the most efficient code.

Divided Differences: We approximate the derivative of F with respect to the ith component of p at a particular point p_0 by either *one-sided differences*

$$\left.\frac{\partial F(p)}{\partial p_i}\right|_{p=p_0} \approx \frac{F(p_0 \pm h * e_i) - F(p_0)}{h} \tag{3}$$

or *central differences*

$$\left.\frac{\partial F(p)}{\partial p_i}\right|_{p=p_0} \approx \frac{F(p_0 + h * e_i) - F(p_0 - h * e_i)}{2h}. \tag{4}$$

Here e_i is the ith Cartesian basis vector. From (1) it can be easily seen that this approach leads to a first- or second-order approximation of the desired derivatives. Divided difference approximations have the advantage that we need only the function as a "black box." A disadvantage, however, is that their accuracy is hard to assess (see, e.g., [13]).

Symbolic Differentiation: Symbolic manipulators like Maple, Macsyma, or Reduce provide powerful capabilities for manipulating algebraic expressions but are, in general, unable to deal with constructs such as branches, loops, or subroutines that are inherent in computer codes. Therefore, differentiation using a symbolic manipulator still requires considerable human effort to break down an existing computer code into pieces digestible by a symbolic manipulator and to reassemble the resulting pieces into a usable derivative code.

Automatic Differentiation: Automatic differentiation techniques rely on the fact that every function, no matter how complicated, is executed on a computer as a (potentially very long) sequence of elementary operations such as additions, multiplications, and elementary functions such as sin and cos (see, for example, [15, 28]). By applying the chain rule

$$\left.\frac{\partial}{\partial t} f(g(t))\right|_{t=t_0} = \left(\left.\frac{\partial}{\partial s} f(s)\right|_{s=g(t_0)}\right) \left(\left.\frac{\partial}{\partial t} g(t)\right|_{t=t_0}\right) \tag{5}$$

over and over again to the composition of those elementary operations, one can compute, in a completely mechanical fashion, derivatives of F that are correct up to machine precision [18]. The techniques of automatic differentiation are directly applicable to computer programs of arbitrary length containing branches, loops, and subroutines. We also note that, unlike handcoding or symbolically assisted approaches, automatic differentiation enables derivatives to be updated easily when the original code changes.

In this paper, we explore different ways of computing first-order derivatives employing automatic differentiation tools. The paper is structured as follows. In the next section we explore the relationship between the tangent linear model and the adjoint approach for computing a gradient from a derivative point of view. In Section 3, we briefly review the forward and reverse mode of automatic differentiation when viewing the automatic differentiation as a code rewriting problem. We then briefly describe the ADIFOR tool for the automatic differentiation of Fortran 77 programs, and present some preliminary results from a sensitivity-enhanced version of MM5 developed with ADIFOR. We then suggest a novel approach to computing large gradients which we call the "pseudoadjoint" approach. It combines what we consider to be the good features of the forward and reverse modes while trying to avoid some of their shortcomings. Lastly, we summarize our results.

2 The Tangent Linear Model and the Adjoint

Two approaches are typically employed in the computation of the linearized model (2)—the tangent linear model (TLM) and the adjoint. They form the basis of the following commonly employed techniques.

Sensitivity analysis techniques — here one tries to asses the sensitivity of the responses of a computational model with respect to perturbations in its parameters or initial conditions (see, for example, [27, 25, 26, 36]).

Data assimilation techniques — here one tries to adjust the initial state of a model to best reproduce some observed behavior (see, for example, [34, 33, 35]).

A collection of papers on this subject can be found in [24].

To illustrate the tangent linear model and the adjoint, we assume that the state X of the system at time t satisfies the simple equation

$$X(t) = H(X(t-1)),\ t = 0, \ldots T \tag{6}$$

and that prognostic and diagnostic variables are the same. Further, let

$$J(i) := \left.\frac{\partial H}{\partial X}\right|_{X=X(i)} \tag{7}$$

be the $n \times n$ Jacobian of H with respect to the state at time step i. The tangent linear model (TLM) describes the linearized evolution of errors about the trajectory of a particular nonlinear solution. Denoting by $\delta X(t)$ the sensitivity of the state at time t with respect to perturbations in the initial state $X(0)$, we have

$$\begin{aligned}
\delta X(1) &= J(0) * \delta X(0) \\
\delta X(2) &= J(1) * \delta X(1) = J(1) * J(0) * \delta X(0) \\
&\vdots \\
\delta X(t) &= J(t-1) * \delta X(t-1) = J(t-1) * \cdots * J(0) * \delta X(0).
\end{aligned} \tag{8}$$

Here $\delta X(t)$ should be interpreted as a column vector. The tangent linear model is defined as

$$X(T) + \delta X(T). \tag{9}$$

Comparing this with (2), we see that the TLM and the first-order Taylor approximation with $\Delta p = \delta X(0)$ are identical. We also point out that by initializing $\delta X_i(0) = 1$, $\delta X_j(0) = 0$ for $j \neq i$, $\delta X(T)$ can be interpreted as the sensitivity of all output variables $X(T)$ with respect to a unit change in the ith component $X_i(0)$ of the initial state. In the literature the expression "development of the TLM" is used somewhat loosely to denote either a code for $\delta X(T)$ or $X(T) + \delta X(T)$ as the computation of $\delta X(i)$ and $X(i)$ is usually intertwined.

In contrast, the adjoint integrates the model back in time. Denote the sensitivity of the final state with respect to a change in an intermediate state by $\overline{\delta X(t)}$. Then

$$\begin{aligned} \overline{\delta X(T-1)} &= \overline{\delta X(T)} * J(T-1) \\ \overline{\delta X(T-2)} &= \overline{\delta X(T-1)} * J(T-2) = \overline{\delta X(T)} * J(T-1) * J(T-2) \\ &\vdots \\ \overline{\delta X(t)} &= \overline{\delta X(T)} * J(T-1) * \cdots * J(t) \end{aligned} \tag{10}$$

Here $\overline{\delta X(t)}$ should be interpreted as a row vector. In particular, initializing $\overline{\delta X_i(T)} = 1$, $\overline{\delta X_j(0)} = 0$ for $j \neq i$, $\overline{\delta X(0)}$ can be interpreted as the sensitivity of the ith component of the final state with respect to a unit change in all components of the initial state. That is, $\overline{\delta X(0)}$ is the gradient

$$\overline{\delta X(0)} \doteq \frac{\partial X_j(T)}{\partial X(0)}$$

and hence can be viewed as another approach for computing the derivatives required for the first-order linear approximation (2).

Let e_i be the ith canonical unit vector, namely, $e_i(i) = 1$, and $e_i(j) = 0$ for $j \neq i$. Then

$$\delta X(0) = e_j \text{ and } \overline{\delta X(T)}^T = e_i$$

implies

$$\forall 1 \leq t \leq T: \ \overline{\delta X(t)}^T * \delta X(t) = e_i^T \frac{\partial X(T)}{\partial X(0)} e_j = \frac{\partial X_i(T)}{\partial X_j(0)}.$$

Hence, combining TLM and adjoint codes, we have many possibilities for computing the same derivative values [8].

To summarize, if we discount numerical instabilities arising from the complementary stability behavior of forward- and backward-integration of dynamical systems (see, for example, [32]), when properly initialized, the tangent linear model and the adjoint will compute the same sensitivities and provide a mechanism for developing a linear approximation of the model. The computational complexity of these two approaches is quite different, though, as we will see in the next section.

3 Automatic Differentiation

Traditionally, two approaches to automatic differentiation (AD) have been developed: the so-called forward and reverse modes. These modes are distinguished by how the chain rule is used to propagate derivatives through the computation. The forward mode accumulates the derivatives of intermediate

```
y(1) = 1.0
y(2) = 1.0
do i = 1,n
   if (x(i) > 0.0) then
      y(1) = y(1) ◇ x(i)
   else
      y(2) = y(2) ◇ x(i)
   endif
enddo
```

Figure 1: Sample Code Fragment

variables with respect to the independent variables, corresponding to the forward sensitivity formalism [11, 12], whereas the reverse mode propagates the derivatives of the final values with respect to intermediate variables corresponding to the adjoint sensitivity formalism [11, 12]. In either case, automatic differentiation produces code that computes the *values* of the analytical derivatives accurate to machine precision.

We illustrate the difference between these two approaches by deriving code for computing $\frac{\partial y}{\partial x(1:n)}$ from the code fragment shown in Figure 1, considering the cases where "◇" is either "*" or "+."

3.1 The Forward Mode

The forward mode of automatic differentiation computes derivatives as shown in Figure 2, much in the way that the chain rule of differential calculus is usually taught. We use the notation ∇**s** to denote the derivative object associated with the program variable **s**. We can easily convince ourselves that by initializing ∇`x(i)` to the ith canonical unit vector of length n, on exit ∇`y(i)` contains the gradient $\frac{\partial y(i)}{\partial x(1:n)}$. In this case, each statement involving a derivative object is really a vector instruction involving n-vectors. On the other hand, if we are interested only in sensitivities with respect to `x(3)`, say, we initialize ∇`x(i) = 0.0` for $i \neq 3$ and ∇`x(3) = 1.0`. In this case, then, each statement involving a derivative object is a scalar instruction, and we emerge with ∇`y(i)` $= \frac{dy(i)}{dx(3)}$. In general, if we view the derivative vectors ∇ as row vectors, the linearity of differentiation implies that the forward mode allows us to compute arbitrary linear combinations of columns of the Jacobian

$$\frac{dy}{dx} = \begin{pmatrix} \frac{\partial y(1)}{\partial x(1)} & \cdots & \frac{\partial y(1)}{\partial x(n)} \\ \frac{\partial y(2)}{\partial x(1)} & \cdots & \frac{\partial y(2)}{\partial x(n)} \end{pmatrix} \tag{11}$$

in that

$$\begin{pmatrix} \nabla y(1) \\ \nabla y(2) \end{pmatrix} = \frac{dy}{dx} * \begin{pmatrix} \nabla x(1) \\ \vdots \\ \nabla x(n) \end{pmatrix}. \tag{12}$$

```
∇y(1) = 0
y(1) = 1.0
∇y(2) = 0
y(2) = 1.0
do i = 1,n
   if (x(i) > 0.0) then
      ∇y(1) = ∇y(1) + ∇x(i)
      y(1) = y(1) + x(i)
   else
      ∇y(2) = ∇y(2) + ∇x(i)
      y(2) = y(2) + x(i)
   endif
enddo
```

Forward Mode for $\diamond = +$

```
∇y(1) = 0
y(1) = 1.0
∇y(2) = 0
y(2) = 1.0
do i = 1,n
   if (x(i) > 0.0) then
      ∇y(1) = x(i)*∇y(1) + y(1)*∇x(i)
      y(1) = y(1) * x(i)
   else
      ∇y(2) = x(i)*∇y(2) + y(2)*∇x(i)
      y(2) = y(2) * x(i)
   endif
enddo
```

Forward Mode for $\diamond = *$

Figure 2: Forward Mode Code for Code Fragment of Figure 1

In particular, initializing `∇x(i) = d(i)`, we compute the directional derivative

$$\frac{dy}{dx} * d = \lim_{h \to 0} \frac{y(x + h * d) - y(x)}{h}. \tag{13}$$

Forward mode code is easy to generate, preserves any parallelizable or vectorizable structures within the original code, and is readily generalized to higher-order derivatives [4]. If we wish to compute m directional derivatives, then running forward-mode code requires at most on the order of m times as much time and memory as the original code.

We also note that what we called the forward mode is the approach often employed when deriving the tangent linear model of a computer code by hand. This should not come as a surprise. Applying the forward mode of automatic differentiation to a code computing (6) will result in the same derivatives as the TLM accumulation (8) when $\nabla X_i(0) = \delta X_i(0)\,, i = 1, \ldots, n$.

3.2 The Reverse Mode

In contrast, the so-called reverse mode of automatic differentiation computes adjoint quantities — the derivative of the final result with respect to an intermediate quantity. To propagate adjoints, we have to be able to reverse the flow of the program, and remember or recompute any intermediate value that nonlinearly impacts the final result.

Let $\overline{s}$ denote the adjoint of a particular variable s. As a consequence of the chain rule it can be shown (see, for example, [18]) that the statement $s = f(v, w)$ in the original code implies

$$\begin{aligned} \overline{v} &\mathrel{+}= \tfrac{\partial s}{\partial v}\overline{s} \\ \overline{w} &\mathrel{+}= \tfrac{\partial s}{\partial w}\overline{s} \end{aligned}$$

```
y(1) = 1.0; y(2) = 1.0;
y1value(0) = y(1); c1 = 0;
y2value(0) = y(2); c2 = 0;
do i = 1,n
   if (x(i) > 0.0) then
      jump(i) = 'left'; c1 = c1 + 1;
      y1value(c1) = y1value(c1 - 1) ◇ x(i)
   else
      jump(i) = 'right'; c2 = c2 + 1;
      y2value(c2) = y2value(c2 - 1) ◇ x(i)
   endif
enddo
```

Figure 3: Code Fragment of Figure 1 Modified in Preparation for Reverse-Mode Code Generation

in the reverse mode code. The notation `a += b` is shorthand for `a = a + b`. When f is a linear elementary operation such as addition or subtraction, $\frac{\partial s}{\partial v} = \frac{\partial s}{\partial w} = 1$, and hence $\frac{\partial s}{\partial v}$ and $\frac{\partial s}{\partial w}$ do not depend on the values of their operands. On the other hand, when f is a nonlinear operation such as a multiplication, both $\frac{\partial s}{\partial v}$ and $\frac{\partial s}{\partial w}$ do depend on the values of their operands, and one must remember either these derivative values or the values of the operands. To be able to reverse the flow of the program, one must also remember intermediate values that were overwritten, and trace how branches were taken. To this end, we transform the code from Figure 1 to the form shown in Figure 3, where we generate a trace of the branch history in the "`jump`" array, and save intermediate values of the variables `y(1)` and `y(2)` in `y1value(:)` and `y2value(:)`. The counters `c1` and `c2` are used to keep track of when a variable was last assigned. We are now in a position to automatically generate reverse mode code for this computation. The result is shown in Figure 4. We use the notation $\overline{\texttt{s}}$ to denote the derivative object associated with the program variable `s`.

We can easily convince ourselves that when we initialize $\overline{\texttt{y(1)}} = 1.0$, $\overline{\texttt{y(2)}} = 0.0$, and all other derivative objects to zero, then by running the codes in Figures 3 and 4, we emerge with $\overline{\texttt{x(i)}} = \frac{\partial x(i)}{\partial y(1)}$. Similarly, initializing $\overline{\texttt{y(1)}} = 0.0$, $\overline{\texttt{y(2)}} = 1.0$, and all other derivative objects to zero, we compute $\overline{\texttt{x(i)}} = \frac{\partial x(i)}{\partial y(2)}$. In general, if we view the adjoint vector associated with a program variable as column vector, the linearity of differentiation implies that

$$\left(\overline{x(1)}, \cdots, \overline{x(n)}\right) = \left(\ \overline{y(1)}\ ,\ \overline{y(2)}\ \right) * \frac{d\,y}{d\,x}, \tag{14}$$

where $\frac{dy}{dx}$ is as defined in equation (11). That is, reverse mode code allows us to compute arbitrary linear combinations of the rows of the Jacobian. Initializing $\overline{\texttt{y(i)}}$ = `d(i)`, we compute the derivative

$$\frac{\partial\,(d^T * y(x))}{\partial\,x}. \tag{15}$$

```
y1value(c1) = y(1); y2value(c2) = y(2);
do i = n to 1 step -1
   if (jump(i) = 'left') then
      y1value(c1 - 1) += yvalue1(c1)
      x(i) += y1value(c1)
      c1 = c1 - 1
   else
      y2value(c2 - 1) += y2value(c2)
      x(i) += y2value(c2)
      c2 = c2 - 1
   endif
enddo
```

Reverse Mode for $\diamond = +$

```
y1value(c1) = y(1); y2value(c2) = y(2);
do i = n to 1 step -1
   if (jump(i) = 'left') then
      y1value(c1 - 1) += x(i)*y1value(c1)
      x(i) += y1value(c1-1)*y1value(c1)
      c1 = c1 - 1
   else
      y2value(c2 - 1) += x(i)*y2value(c2)
      x(i) += y2value(c2-1)*y2value(c2)
      c2 = c2 - 1
   endif
enddo
```

Reverse Mode for $\diamond = *$

Figure 4: Reverse Mode Code Generated from Code in Figure 3

Note that it is quite an involved process to generate reverse mode code. While the complexity of the forward-mode code generation in Figure 2 changed minimally when we considered an addition instead of a multiplication, the reverse mode code is very sensitive to this change: there is no need to save the intermediate values of `y(1)` and `y(2)` when $\diamond = +$, but we must save them when $\diamond = *$, at the expense of an extra $O(n)$ memory locations. Extra storage is required to remember the way the branches were taken, regardless of whether the loop computed a multiplication or addition. Hence, the reverse mode can, in extreme cases, require as much memory for the tracing of intermediate values and branches as there are floating-point operations being executed during the run of the program. However, its running time is roughly m times that of the function when computing m linear combinations of the rows of the Jacobian. This is particularly advantageous for gradients, since then $m = 1$.

We also note that what we called the reverse mode of automatic differentiation is what is usually performed by hand when deriving an adjoint code. Applying the reverse mode of automatic differentiation to a code computing (6) will result in the same derivatives as the adjoint accumulation (10) when $\overline{X_i(T)} = \overline{\delta X_i(T)}, i = 1, \ldots, n$.

4 The ADIFOR Automatic Differentiation Tool and an Application to the MM5 Mesoscale Weather Model

There have been various implementations of automatic differentiation; an extensive survey can be found in [22]. In particular, we mention GRESS [20], and PADRE-2 [23] for Fortran Programs and ADOL-C [17] for C programs. GRESS, PADRE-2, and ADOL-C implement both the forward and reverse mode. To save control flow information and intermediate values, these tools generate a "trace" of the computation by writing down the particulars of every operation performed in the code. The interpretation overhead associated with using this trace for the purposes of automatic differentiation and its potentially very large size can be a serious computational bottleneck [31].

4.1 The ADIFOR (Automatic Differentiation of Fortran) Tool

Recently, a "source transformation" approach to automatic differentiation has been explored in the ADIFOR [2], ADIC [7], and Odyssee [29, 30] tools. ADIFOR and Odyssee transform Fortran 77 code and ADIC transforms ANSI-C code. By applying the rules of automatic differentiation, these tools generate new code, which, when executed, computes derivatives without the overhead associated with trace interpretation schemes. ADIFOR and ADIC mainly use the forward mode[2]. In contrast, Odyssee employs the reverse mode.

Given a Fortran subroutine (or collection of subroutines) describing a "function," and an indication of which variables in parameter lists or common blocks correspond to "independent" and "dependent" variables with respect to differentiation, ADIFOR performs a data flow analysis to determine which statements in the code have to be augmented with derivative computations and then produces Fortran 77 code that computes the derivatives of the dependent variables with respect to the independent ones. ADIFOR produces portable Fortran 77 code, and accepts almost all of Fortran 77; in particular, it can deal with arbitrary calling sequences, nested subroutines, common blocks, and equivalences. The ADIFOR-generated code tries to preserve vectorization and parallelism in the original code and employs a consistent subroutine-naming scheme, which allows for code tuning, the exploitation of domain-specific knowledge, and the use of vendor-supplied libraries.

ADIFOR employs a hybrid forward/reverse mode approach to generating derivatives. For each assignment statement, it uses the reverse mode to generate code that computes the partial derivatives of the result with respect to the variables on the right-hand side and then employs the forward mode to propagate overall derivatives. For example, the single Fortran statement

$$\mathtt{y} = \mathtt{x(1)} * \mathtt{x(2)} * \mathtt{x(3)} * \mathtt{x(4)} * \mathtt{x(5)}$$

gets transformed into the code segment shown in Figure 5. Note that none of the common subexpressions $x(i) * x(j)$ are recomputed in the reverse mode section for $\frac{\partial \mathtt{y}}{\partial \mathtt{x(i)}}$. The variable **g$p$** denotes the number of (directional) derivatives being computed. For example, if **gp** $= 5$, and **g$x(1:5,1:5)** is the 5×5 identity matrix (i.e., **g$x(i,j)** $= \frac{\partial x(i)}{\partial x(j)}$) then upon execution of these statements, **g$y(1:5)** equals $\frac{dy}{dx}$. On the other hand, assume that we wished only to compute derivatives with respect to a scalar parameter `s`, so **gp** $= 1$, and, on entry to this code segment, **g$x(1,i)** $= \frac{\partial x(i)}{\partial s}$, $i = 1, \ldots, 5$. Then the do-loop in Figure 5 implicitly computes $\frac{dy}{ds} = \frac{dy}{dx}\frac{dx}{ds}$ without ever forming $\frac{\partial y}{\partial x}$ explicitly. Note that the cost of computing y is amortized over all the derivatives being computed. Thus, this approach is more efficient than the normal forward mode or a divided-difference approximation when more than a few derivatives are computed at the same time.

We also see that ADIFOR-generated code provides the directional derivative computation possibilities associated with the forward mode of automatic differentiation [6]. Instead of simply producing code to compute the Jacobian J, ADIFOR produces code to compute $J * S$, where the "seed matrix" S is

[2]Information on ADIFOR and ADIC can be found on the world-wide web under **http://www.mcs.anl.gov/Projects/autodiff/index.html**

```
r$1 = x(1) * x(2)
r$2 = r$1 * x(3)
r$3 = r$2 * x(4)
r$4 = x(5) * x(4)
r$5 = r$4 * x(3)
r$1bar = r$5 * x(2)
r$2bar = r$5 * x(1)
r$3bar = r$4 * r$1
r$4bar = x(5) * r$2
```

Reverse Mode for computing $\frac{\partial y}{\partial x(i)}$:

$$r\$jbar = \frac{\partial y}{\partial x(i)},\ i = 1, \ldots, 4$$

$$r\$3 = \frac{\partial y}{\partial x(5)}$$

```
do g$i$ = 1, g$p$
   g$y(g$i$) = r$1bar * g$x(g$i$,1)
             + r$2bar * g$x(g$i$,2)
             + r$3bar * g$x(g$i$,3)
             + r$4bar * g$x(g$i$,4)
             + r$3 * g$x(g$i$, 5)
enddo
```

Forward Mode:
Assembling ∇y from $\frac{\partial y}{\partial x(i)}$ and $\nabla x(i)$, $i = 1, \ldots, 5$.

```
y = r$3 * x(5)
```

} Computing function value

Figure 5: Sample Segment of an ADIFOR-generated Code

initialized by the user. Thus, if S is the identity, ADIFOR computes the full Jacobian; whereas if S is just a vector, ADIFOR computes the product of the Jacobian by a vector. In [5] the flexibility of the ADIFOR interface is exploited in a "stripmining" approach to decrease turnaround time for derivative computations by spawning several independent subprocesses computing parts of the desired gradient or Jacobian. The seed matrix also provides a powerful mechanism for decreasing the computational complexity of derivative codes through judicious use of the chain rule [8, 21]. The running time and storage requirements of the ADIFOR-generated code are roughly proportional to the numbers of columns of S, which equals the **gp** variable in the sample code above.

ADIFOR has been successfully applied to codes from various domains of science. Experiences with meteorological codes, for example, have been reported in [10, 25, 26, 27]. Typically, ADIFOR-generated code runs two to four times faster than one-sided divided difference approximations when one computes more than 5-10 derivatives at one time. The superior performance is due to the reverse/forward hybrid mode and a dependence analysis that tries to avoid computing derivatives of expressions that do not affect the "dependent variables". We also note that in order to take full advantage of reduced complexity of ADIFOR-generated code, it is advantageous to compute several directional derivatives at the same time — so the ADIFOR-generated code may require significantly more memory than the original simulation code.

4.2 First Results with a Sensitivity-Enhanced Version of the MM5 Mesoscale Weather Model

The development of a sensitivity-enhanced version of the Fifth-Generation Penn State/NCAR mesoscale weather model (MM5) [14] using the ADIFOR automatic differentiation tool is in progress. ADIFOR expects code that complies with the Fortran 77 standard. MM5 does not comply with this standard —

in particular, it makes much use of "pointer variables." We circumvented this difficulty by developing an MM5-specific tool to map the pointer handling to standard-conforming Fortran77 code acceptable to ADIFOR, and to remap ADIFOR's output to obtain the desired sensitivity-enhanced code.

To date, we have developed sensitivity-enhanced versions of the nonhydrostatic dynamics and most of the physics, including the Blackadar high-resolution planetary boundary layer parameterization, the Grell and Kuo cumulus parametrizations, the Dudhia long- and short-wave radiation scheme, and the models representing explicit moisture with treatment of mixed phase processes (ice), shallow convection, and dry convective adjustment.

To verify our ADIFOR-generated sensitivities, we used the sensitivity-enhanced version of MM5 to compute the time-evolution of first-order perturbations of pressure, temperature, water vapor, and convective rain, in response to perturbations in the initial-pressure data. Our perturbation scheme introduces an artificial perturbation parameter ϵ. To properly address the leapfrog scheme employed in MM5, we perturbed the pressure[3] on the two initial time-slices t_0 and t_1 as follows:

$$\begin{array}{rcl} p'(x,y,z,t_0;\,\epsilon) & := & (1+\epsilon)\,p'(x,y,z,t_0) \\ p'(x,y,z,t_1;\,\epsilon) & := & (1+\epsilon)\,p'(x,y,z,t_1) \end{array} . \tag{16}$$

Then ϵ parameterizes a family of perturbed solutions for the MM5 variables, with the unperturbed solution obtained at $\epsilon = 0$. Any quantity (Q, say) influenced by the initial pressure therefore acquires an implicit dependence on ϵ. Pressure, temperature, water vapor, and convective rain, are particular instances of such quantities. By definition, the first-order perturbation-theoretic sensitivity of Q is

$$\delta Q(x,y,z,t) := \left. \frac{\partial\, Q(x,y,z,t;\,\epsilon)}{\partial\,\epsilon} \right|_{\epsilon=0} . \tag{17}$$

Given our choice of the perturbation of p', $\delta Q(x,y,z,t)$ can be interpreted as the sensitivity of Q to a uniform relative change in the initial pressure fields. We note that both $p'(t_0)$ and $p'(t_1)$ were perturbed, since a leapfrog timestepping approach is employed in MM5.

The sensitivity $\delta Q(x,y,z,t)$ can easily be computed by using automatic differentiation. We slightly modify MM5 to include ϵ as an input parameter entering into the computation of the initial pressre, as shown above. We then employ an automatic differentiation tool (ADIFOR, in our case) to differentiate the code with respect to ϵ. Finally, we evaluate the sensitivity-enhanced code at $\epsilon = 0$. In this fashion, we can compute quantities equivalent to those obtained in the tangent linear model *without any further hand-modification of the code.*

We compared the results of our perturbation AD-based method with the results of perturbing the initial-pressure data by one part in 10^3 in the fully nonlinear model, which corresponds to choosing $\epsilon = \pm 10^{-3}$ in Equation (16). For each of pressure (MM5 variable **ppa**), temperature (**ta**), water vapor (**qva**), and convective rain (**rainc**), we computed the following quantities:

- Its value using the initial pressure distribution.
- Its sensitivity $\delta Q(x,y,z,t)$ provided by the ADIFOR-generated code.

[3]Following standard notation, $p' = p - p_0$ is the deviation of the absolute pressure p from the reference-state pressure p_0; this is the quantity actually computed by MM5 in nonhydrostatic mode.

- An approximation to this sensitivity via a central divided-difference approximation (4). Hence, for example, choosing $Q = \mathbf{ppa}$, we also compute the values $\mathbf{ppa}_\pm$ obtained by a relative perturbation of the initial pressure distribution by $\pm\epsilon = 10^{-3}$, as in Equation (16). We then computed the second-order divided difference approximation $\tilde{\delta}\mathbf{ppa}$ to $\delta\mathbf{ppa}$:

$$\tilde{\delta}\mathbf{ppa}(x, y, z, t, \epsilon) := \frac{\mathbf{ppa}_+ - \mathbf{ppa}_-}{2\epsilon}.$$

- The difference

$$\Delta Q := \tilde{\delta}Q - \delta Q$$

between the divided-difference approximation and the first-order sensitivity. It follows from (1) that, assuming that the model is continuously differentiable in the vicinity of the base state, ΔQ should be of the order $O(\epsilon^3)$ and hence negligible except for areas of high nonlinearity. It should be noted, however, that there is no guarantee that the model is continuously differentiable—in particular, in light of the switching behavior in the moisture physics modeling.

We note that for this particular case the ADIFOR-generated code exhibits run times and storage requirements on the order of twice those of the original MM5 code, and hence it obtains first-order sensitivities at about two-thirds the cost of the divided-difference approximation.

The data set we had at our disposal for testing purposes uses a $28 \times 25 \times 23$ grid, without nesting. We ran the model for 12 time steps of 4 minutes each, terminating at 48 minutes. In Figure 6, for the final time-step, we show for surface pressure

- the value of the forecast variable,
- the sensitivity of the forecast variable with respect to a uniform relative change of the pressure at time zero, and
- the difference between the first-order sensitivity and the divided-difference approximation.

We mention that this perturbation may violate some consistency conditions enforced by the MM5 pre-processors. We would have had to perturb the vertical wind velocity as well, in order not to excite sound waves. This issue will also be addressed by generating sensitivity-enhanced versions of the MM5 pre-processors. We did observe these waves, but they quickly propagate out of the system. At any rate, the purpose of this exercise was to validate our ADIFOR-generated derivatives. Since pressure couples with all the other variables, it still is an appropriate choice to check the correctness of the ADIFOR-generated derivative propagation code.

The upper left plot in Figure 6 shows the value of $\mathbf{ppa}$ after the twelfth time step, the upper right plot shows the sensitivity $\delta\mathbf{ppa}$, and the bottom plots shows $\Delta\mathbf{ppa}$. Pressure is in Pascals, and $\delta\mathbf{ppa}$ and $\Delta\mathbf{ppa}$ are in Pascals per unit ϵ. The labels in the legend denote the boundaries between contour intervals.

As we can see, excellent first-order agreement is achieved everywhere for surface pressure. We see that the fractional differences between the first-order sensitivity and its divided-difference approximation are only a few parts in 10^3. Since the perturbation is the sensitivity times ϵ, it follows that the absolute differences between the tangent-linear and fully nonlinear models of the surface-pressure field are only of the order of a few parts in 10^6.

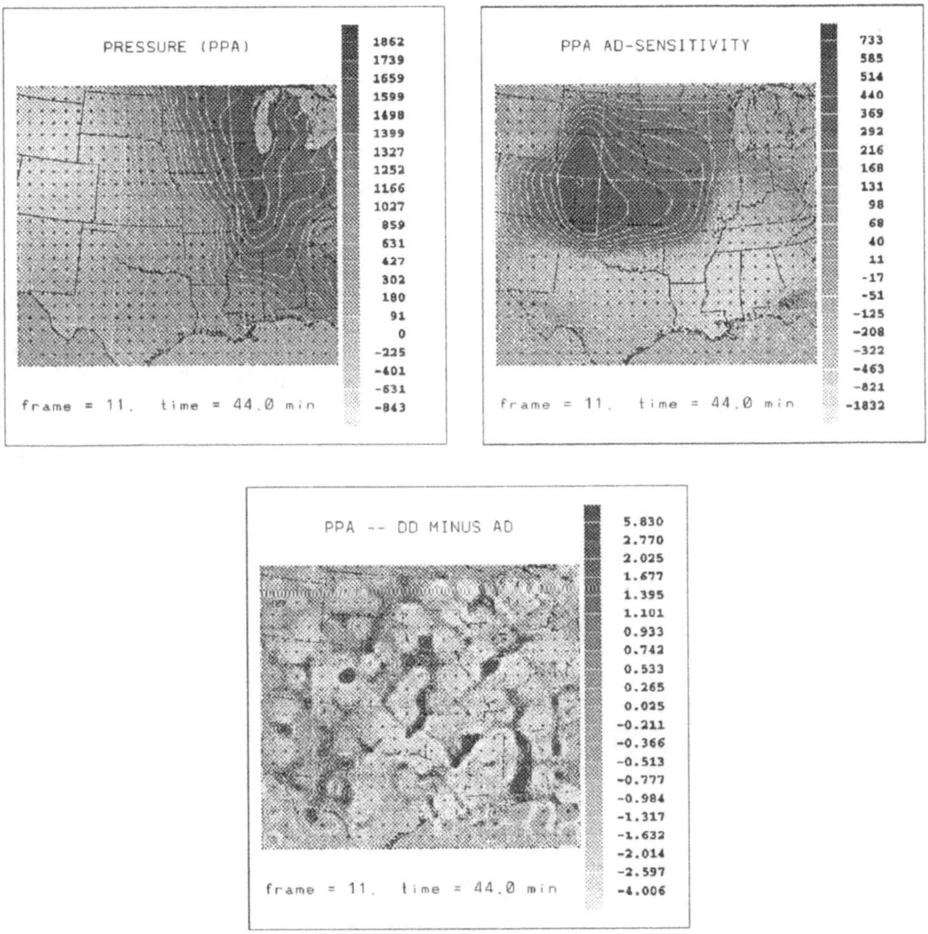

Figure 6: Plot showing the surface-pressure field **ppa** (upper left), the first-order sensitivity δ**ppa** (upper right), and the difference Δ**ppa** (bottom)

5 Pseaudoadjoints

Let us now assume that we have a computation where we repeatedly update a (large) state X and at the end use a merit function that summarizes the relevant features of the final state in a few numbers. That is, we can view the computation as follows.

$$X(0) \xrightarrow{H} X(1) \xrightarrow{H} \cdots \xrightarrow{H} X(T) \xrightarrow{R} r, \tag{18}$$

where, as in (6), H denotes the update operator and R denotes the merit function. We further assume for simplicity that $1 = \dim(r) \ll n := \dim(X)$.

We are interested in efficiently evaluating the gradient

$$\frac{\partial r}{\partial X(0)} = \frac{\partial r}{\partial X(T)} * \frac{\partial X(T)}{\partial X(T-1)} * \ldots * \frac{\partial X(1)}{\partial X(0)}. \tag{19}$$

A forward-mode based approach like that used in ADIFOR will require a run time that is proportional to n, whereas a reverse mode approach might require a run time that is comparable to a few function evaluations. Of course, the complexity of the reverse mode approach greatly depends on the operator H. If H is "mostly linear," then applying the reverse mode to H does not require much storage for intermediate values. A "highly nonlinear" H, on the other hand, may pose considerable storage demands.

Let us further assume that

- $\frac{\partial R}{\partial x}$ can be computed easily (often it is a weighted sum of squares) and that
- $\frac{\partial H}{\partial X}$ is sparse. This situation is fairly typical because of the local nature of stencil-based approximations.

5.1 Exploiting Sparsity in Forward-Mode Derivative Computations

Forward-mode approaches can be used advantageously to compute large sparse Jacobians. One can employ a so-called compressed Jacobian approach, which, given the sparsity pattern of the Jacobian, derives a graph coloring that identifies which columns of the Jacobian can be computed with the same directional derivative. To illustrate, let us assume that we have a function

$$F = \begin{pmatrix} f_1 \\ f_2 \\ f_3 \\ f_4 \\ f_5 \end{pmatrix} : x \in \mathbf{R}^4 \mapsto y \in \mathbf{R}^5$$

whose Jacobian J has the following structure (symbols denote nonzeros, and zeros are not shown):

$$J = \begin{pmatrix} \bigcirc & & & \\ \bigcirc & & & \Diamond \\ & \triangle & & \Diamond \\ & \triangle & \Box & \\ & \triangle & \Box & \end{pmatrix}.$$

Columns 1 and 2, as well as columns 3 and 4, are structurally orthogonal. In divided-difference approximations one could exploit that structure by perturbing both x_1 and x_2 in one function evaluation, and both x_3 and x_4 in the other. In an automatic differentiation system like ADIFOR, one can exploit this fact by setting

$$S = \begin{pmatrix} 1 & 0 \\ 1 & 0 \\ 0 & 1 \\ 0 & 1 \end{pmatrix},$$

which results in the computation of the compressed Jacobian

$$JS = \begin{pmatrix} \bigcirc & \\ \bigcirc & \diamond \\ \triangle & \diamond \\ \triangle & \square \\ \triangle & \square \end{pmatrix}$$

at roughly half the cost. For most grid problems, the width p of the compressed Jacobian is independent of the problem size and depends only on the local stencil chosen. Experimental results with this approach in computing large sparse Jacobians as they arise in large-scale nonlinear equations have been reported in [1].

This compressed Jacobian approach is also applicable to the computation of gradients of so-called partially separable functions [19], which are functions f that can be represented in the form

$$f(x) = \sum_{i=1}^{np} f_i(x), \tag{20}$$

where each of the component functions f_i has limited support. Hence, the gradients ∇f_i are sparse, even though the final gradient ∇f is dense. It can be shown [19] that any function with a sparse Hessian is a partially separable one. The computation of the gradient of a partially separable function can be reduced to the problem of computing a sparse Jacobian [9] by realizing that the gradient of f can easily be obtained by summing the rows of the *sparse* Jacobian $\frac{dG}{dx}$, where

$$G(x) = \begin{pmatrix} f_1(x) \\ \vdots \\ f_{np}(x) \end{pmatrix}. \tag{21}$$

Another approach is based on the realization that the workhorse of any mainly forward-mode first-order automatic differentiation approach is a "vector linear combination," for example, `∇y(1) = x(i)*∇y(1) + y(1)*∇x(i)` in Figure 3. Here `∇y(1)` and `∇x(i)` are vectors of length p, where, as in Subsection 4, p denotes the number of directional derivatives to be computed, and `y(1)` and `x(i)` are scalars. This operation is a particular instantiation of

$$w = \sum_{i=1}^{k} \alpha_i * v_i, \tag{22}$$

where w and v_i are vectors of length p, the α_i are scalar multipliers, and k is referred to as the "arity."

If the initial seed matrix is sparse (e.g., the identity), then, if we ignore exact numerical cancellation, the left-hand side vector w in (22) has no fewer nonzeros than any of the vectors on the right-hand side. Hence, if the final derivative objects, which correspond to a row of the Jacobian J or a component gradient ∇f_i, are sparse, all intermediate vectors must be sparse. That is, by expressing the derivative linear combinations with algorithms and data structures tailored toward to sparse vectors, we can exploit sparsity in a transparent fashion, *even if the sparsity pattern of the derivative matrix is not known beforehand.* Also note that the sparsity structure of J or ∇f_i is computed as a byproduct of the derivative computation.

The SparsLinC (Sparse Linear Combination) Library [3, 8] addresses the scenario where p is large and most of the vectors involved in vector linear combination are sparse. It provides support for sparse vector linear combination, in a fashion that is well suited to the use of this operation in the context of automatic differentiation. SparsLinC, which is written in ANSI C, includes the following features:

Three data structures for sparse vectors: SparsLinC has different data structures for a vector containing only one nonzero, a few nonzeros, or several nonzeros.

Efficient Memory Allocation Scheme: SparsLinC employs a "bucket" memory allocation scheme which, in effect, provides a buffered memory allocation mechanism.

Polyalgorithms: SparsLinC switches between vector representations in a transparent fashion and provides special support for the "+=" operation $w = \alpha_1 * w + \alpha_2 * v$, which occurs frequently when computing gradients of partially separable functions, as suggested by (20).

Full Precision Support: single and double precision computations are provided for both real- and complex-valued computations.

In this fashion, SparsLinC can adapt to the dynamic nature of the derivative vectors, efficiently representing derivative vectors that grow from a column of the identity matrix (often occurring in the ADIFOR seed matrix) to a dense vector, such as ∇f in (20). We also mention that almost no memory is allocated for derivative objects that are all zeros. SparsLinC will be fully integrated in the forthcoming release of the ADIFOR 2.0 system.

Note that neither the run time of the compressed Jacobian nor that of the SparsLinC approach is affected by nonlinearities in the program. That is, changing all additions in the code to multiplications does not increase the required storage, and the run time increases at most by a factor of two.

5.2 Pseudoadjoints

Now, coming back to the problem of computing $\frac{\partial r}{\partial X(0)}$, we realize that if we consider the computation of $\frac{\partial H}{\partial X}$ as a "black box," we can interpret equation (19) as a series of matrix-matrix and vector-matrix multiplications, namely,

where the vector on the left-hand side corresponds to $\frac{\partial r}{\partial X(0)}$, the vector on the right-hand side corresponds to $\frac{\partial r}{\partial X}|_{X=X(T)}$, and the matrices on the right-hand side correspond to

$$J(i) := \frac{\partial H}{\partial X}|_{X=X(i)}, i = 1, \ldots, T-1.$$

An automatic differentiation tool that employs the forward mode throughout can be interpreted as accumulating this product from the right, forming a large matrix corresponding to $\frac{\partial X(T)}{\partial X(0)}$, which then is multiplied by the vector $\frac{\partial r}{\partial X(T)}$ to result in the final gradient. Obviously, it would be much more advantageous to accumulate the product from the left, computing $\frac{\partial r}{\partial X(i)}, i = T, \ldots, 1$ through a series of matrix-vector multiplies. The first approach can be interpreted as the forward mode in an algebra where H and R are elementary operators, the latter one as the reverse mode in this algebra.

However, using the approaches outlined in the previous subsection, we are in a position to inexpensively compute $\frac{\partial H}{\partial X}$, with predictable storage and runtime requirements that do not depend on the nonlinearity of the model. Hence, by storing all intermediate derivatives $J(i)$ in a forward sweep. we can compute $\frac{dr}{dX(0)}$ through a series of sparse matrix-vector multiplies at a runtime complexity that is of the order

$$T * \text{runtime}(\frac{\partial H}{\partial X}) + T * \text{runtime(sparse matrix-vector multiply)}$$

and the storage complexity is of the order

$$T * \text{storage}(\frac{\partial H}{\partial X}).$$

If we only store snapshots of X in the forward pass and regenerate individual Jacobians $\frac{\partial H}{\partial X}|_{X=X(i)}$ when needed, the runtime complexity is of the order

$$T * \text{runtime}(\frac{\partial H}{\partial X}) + T * \text{runtime}(H) + T * \text{runtime(sparse matrix-vector multiply)}$$

and storage complexity that is of the order

$$\text{storage}(\frac{\partial H}{\partial X}) + T * \text{storage}(X).$$

When we employ the snapshotting scheme proposed by Griewank [16] to regenerate the $X(i)$ from a series of checkpoints, the time complexity of the latter approach becomes of the order

$$\begin{aligned} T * \text{runtime}(\frac{\partial H}{\partial X}) \quad &+ \quad T * (1 + \log(T)) * \text{runtime}(H) \\ &+ \quad T * \text{runtime(sparse matrix-vector multiply)} \end{aligned}$$

and storage complexity is of the order

$$\text{storage}(\frac{\partial H}{\partial X}) + (1 + \log(T)) * \text{storage}(X).$$

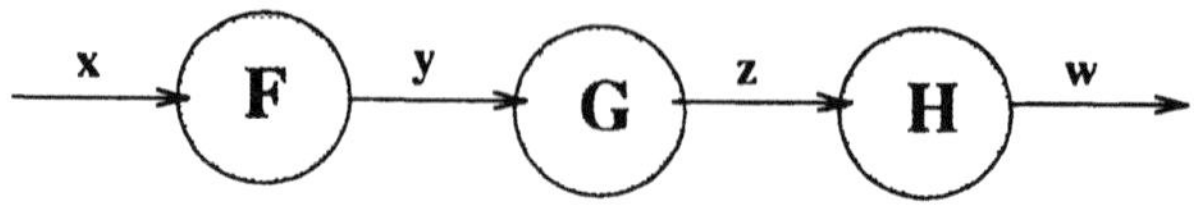

Figure 7: Serial Simulation

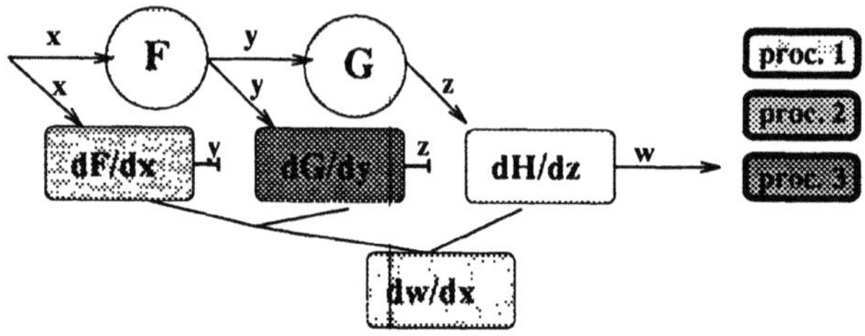

Figure 8: Parallel Derivative Computation

Given that we can compute sparse Jacobians efficiently, at a storage and memory cost that is a moderate multiple of the cost for evaluating H itself, the last approach may be an attractive approach to developing a gradient code. It shares with a "usual" adjoint the characteristic that the ratio of gradient to function evaluation does not depend on the number of input parameters. This approach capitalizes on the advantages of the forward mode (efficient computation of sparse Jacobians with predictable complexity independent of the nonlinearity of the model) and the reverse mode (lower arithmetic complexity). It also avoids the drawbacks of the forward mode (the ratio of gradient to function evaluation depends on the number of input variables) and the reverse mode (a highly nonlinear model may lead to excessive storage demands).

However, if execution time is of the essence, we can do even better by exploiting chain rule associativity to break the time dependency in the derivative computation. To illustrate this idea, we consider the situation shown in Figure 7: G cannot start before F has been computed, and H has to wait for the completion of G. That is, none of these processes may execute in parallel. This is not the case in derivative computations, however, because of the associativity of the derivative chain rule. For example, we could proceed as in Figure 8. That is, at the same time that we spawn a process to compute F, we spawn a process to compute $\frac{dy}{dx}$, and at the same time that we start with computation of G, we spawn a process to compute $\frac{dz}{dy}$. Lastly, the computation of $\frac{dw}{dz}$ is initiated. Under the assumption that the computation of derivatives takes significantly longer than the simulation itself, we will, in the end, have the three derivative processes running in parallel When they have finished, we simply accumulate their outputs to arrive at the desired result, $\frac{dw}{dx}$. Thus, if we are willing to duplicate the computation of y and z, we can in this fashion arrive at a coarse-grained parallel schedule that, with minimal synchronization requirements, could be mapped to a network of workstations.

In particular, if we now apply this idea to our problem (18), we could, if enough memory and processors were available, spawn processes to independently compute $J(i)$ in parallel, and would only have to wait for an *additional* run time of the order of

$$\text{runtime}(\frac{\partial H}{\partial X}) + T * \text{runtime(sparse matrix-vector multiply)},$$

to obtain the desired gradient $\frac{\partial r}{\partial X(0)}$.

6 Conclusions and Future Work

This paper gave a brief introduction into automatic differentiation and related it to the tangent linear model and adjoint approaches commonly used in meteorology. We briefly reviewed the forward and reverse mode of automatic differentiation and introduced the ADIFOR automatic differentiation tool. We also presented first results of a sensitivity-enhanced version of the MM5 PSU/NCAR mesoscale weather model, thus demonstrating that automatic differentiation can generate results equivalent to a tangent linear model for sophisticated weather models, with minimal recourse to laborious and error-prone hand-coding.

We presented a novel approach to the computation of gradients, which used a reverse mode approach at the timestep level and a forward mode approach at every time step. The resulting "pseudoadjoint" shared the characteristic of an adjoint code that the ratio of gradient to function evaluation did not depend on the number of independent variables, but, in contrast to a true adjoint approach, the nonlinearity of the model played no role in the complexity of the derivative code. Lastly, we motivated how chain rule associativity could be employed to break time dependencies in the derivative computation.

The pseudoadjoint strategy is a particular instantiation of what we call a "hybrid mode," where both forward and reverse modes of automatic differentiation are employed at various levels in the derivative computation. The strategy employed in the ADIFOR and ADIC tools is a particularly simple instance of such a hybrid strategy which, in some sense, is at the opposite spectrum of the "pseudoadjoint" approach suggested here. In the pseudoadjoint approach suggested in Section 5 we are using the reverse mode at the outermost loop level, and a forward based technique in the rest, whereas ADIFOR and ADIC employs the reverse mode at the lowest level, within the scope of an assignment statement. Clearly, there are many other alternatives, and we are beginning to explore them in a systematic fashion with an eye toward capitalizing on the benefits of the forward and reverse mode approaches while avoiding their respective drawbacks.

Acknowledgments

We thank Andreas Griewank for many stimulating discussions on the subject, Alan Carle for his instrumental role in the ADIFOR project, and Gordon Pusch for his essential role in the development of the sensitivity-enhanced version of the MM5 code. We also thank Kelvin Droegemeier and Seon Ki Park for introducing us to the use of tangent linear and adjoint models in meteorology.

References

[1] Brett Averick, Jorge Moré, Christian Bischof, Alan Carle, and Andreas Griewank. Computing large sparse Jacobian matrices using automatic differentiation. *SIAM Journal on Scientific Computing*, 15(2):285–294, 1994.

[2] Christian Bischof, Alan Carle, George Corliss, Andreas Griewank, and Paul Hovland. ADIFOR: Generating derivative codes from Fortran programs. *Scientific Programming*, 1(1):11–29, 1992.

[3] Christian Bischof, Alan Carle, and Peyvand Khademi. Fortran 77 interface specification to the SparsLinC library. Technical Report ANL/MCS-TM-196, Mathematics and Computer Science Division, Argonne National Laboratory, 1994.

[4] Christian Bischof, George Corliss, and Andreas Griewank. Computing second- and higher-order derivatives through univariate Taylor series. *Optimization Methods and Software*, 2:211–232, 1993.

[5] Christian Bischof, Larry Green, Kitty Haigler, and Tim Knauff. Parallel calculation of sensitivity derivatives for aircraft design using automatic differentiation. In *Proceedings of the 5th AIAA/NASA/USAF/ISSMO Symposium on Multidisciplinary Analysis and Optimization*, pages 73–84. American Institute of Aeronautics and Astronautics, 1994. AIAA 94-4261.

[6] Christian Bischof and Paul Hovland. Using ADIFOR to compute dense and sparse Jacobians. Technical Report ANL/MCS-TM-158, Mathematics and Computer Science Division, Argonne National Laboratory, 1991.

[7] Christian Bischof and Andrew Mauer. ADIC – a tool for the automatic differentiation of C programs. Unpublished Information, Argonne National Laboratory, 1994.

[8] Christian H. Bischof. Automatic differentiation, tangent linear models and pseudo-adjoints. Preprint MCS-P472-1094, Mathematics and Computer Science Division, Argonne National Laboratory, 1994.

[9] Christian H. Bischof and Moe El-Khadiri. Extending compile-time reverse mode and exploiting partial separability in ADIFOR. Technical Report ANL/MCS-TM-163, Mathematics and Computer Science Division, Argonne National Laboratory, 1992.

[10] Daewon W. Byun, Robert Dennis, Dongming Hwang, Jr. Carlie Coats, and M. Talat Odman. Computational modelling issues in next generation air quality models. In *Proceedings of IMACS'94, Atlanta, Georgia*, 1994.

[11] D. G. Cacuci. Sensitivity theory for nonlinear systems, I: Nonlinear functional analysis approach. *Journal of Mathematical Physics*, 22(12):2794–2802, 1981.

[12] D. G. Cacuci. Sensitivity theory for nonlinear systems, II: Extension to additional classes of responses. *Journal of Mathematical Physics*, 22(12):2803–2812, 1981.

[13] Phillip E. Gill, Walter Murray, and Margaret H. Wright. *Practical Optimization*. Academic Press, London, 1981.

[14] G. A. Grell, J. Dudhia, and D. R. Stauffer. A description of the fifth-generation Penn State/NCAR mesoscale weather model (MM5). Technical Report NCAR/TN-398+STR, National Center for Atmospheric Research, Boulder, Colorado, June 1994.

[15] Andreas Griewank. The chain rule revisited in scientific computing. Preprint MCS–P227–0491, Mathematics and Computer Science Division, Argonne National Laboratory, 1991.

[16] Andreas Griewank. Achieving logarithmic growth of temporal and spatial complexity in reverse automatic differentiation. *Optimization Methods and Software*, 1(1):35–54, 1992.

[17] Andreas Griewank, David Juedes, and Jay Srinivasan. ADOL-C, a package for the automatic differentiation of algorithms written in C/C++. Preprint MCS-P180-1190, Mathematics and Computer Science Division, Argonne National Laboratory, 1990.

[18] Andreas Griewank and Shawn Reese. On the calculation of Jacobian matrices by the Markowitz rule. In Andreas Griewank and George F. Corliss, editors, *Automatic Differentiation of Algorithms: Theory, Implementation, and Application*, pages 126–135. SIAM, Philadelphia, 1991.

[19] Andreas Griewank and Philippe L. Toint. On the unconstrained optimization of partially separable objective functions. In M.J.D. Powell, editor, *Nonlinear Optimization 1981*, pages 301–312, London, 1981. Academic Press.

[20] Jim E. Horwedel. GRESS: A preprocessor for sensitivity studies on Fortran programs. In Andreas Griewank and George F. Corliss, editors, *Automatic Differentiation of Algorithms: Theory, Implementation, and Application*, pages 243–250. SIAM, Philadelphia, 1991.

[21] Paul Hovland, Christian Bischof, Donna Spiegelman, and Mario Casella. Efficient derivative codes through automatic differentiation and interface contraction and an application in biostatistics. in preparation.

[22] David Juedes. A taxonomy of automatic differentiation tools. In Andreas Griewank and George Corliss, editors, *Proceedings of the Workshop on Automatic Differentiation of Algorithms: Theory, Implementation, and Application*, pages 315–330, Philadelphia, 1991. SIAM.

[23] Koichi Kubota. PADRE2, a FORTRAN precompiler yielding error estimates and second derivatives. In Andreas Griewank and George F. Corliss, editors, *Automatic Differentiation of Algorithms: Theory, Implementation, and Application*, pages 251–262. SIAM, Philadelphia, 1991.

[24] Special Issue on Adjoint Applications in Dynamic Meterology. Tellus, 45a(5), 1993.

[25] Seon Ki Park and Kelvin Droegemeier. Effect of a microphysical parameterization on the evolution of linear perturbations in a convective cloud model. In *Preprints, Conference on Cloud Physics, January 1995, Dallas, Texas.* American Meteorological Society.

[26] Seon Ki Park and Kelvin Droegemeier. On the use of automatic differentiation to assess parametric sensitivity in convective-scale variational data assimilation. In *Preprints, Proc. Int. Symp. on Assimilation of Observations in Meteorology and Oceanography, Tokyo, Japan, March 1995.* World Meteorological Organization.

[27] Seon Ki Park, Kelvin Droegemeier, Christian Bischof, and Tim Knauff. Sensitivity analysis of numerically-simulated convective storms using direct and adjoint methods. In *Preprints, 10th Conference on Numerical Weather Prediction, Portland, Oregon*, pages 457–459. American Meterological Society, 1994.

[28] Louis B. Rall. *Automatic Differentiation: Techniques and Applications*, volume 120 of *Lecture Notes in Computer Science*. Springer Verlag, Berlin, 1981.

[29] Nicole Rostaing, Stephane Dalmas, and Andre Galligo. Automatic differentiation in Odysee. *Tellus*, 45a(5):558–568, October 1993.

[30] Nicole Rostaing-Schmidt and Eric Hassold. Basic functional representation of programs for automatic differentiation in the Odyssee system. In Francois-Xavier Le Dimet, editor, *High-Performance Computing in the Geosciences*, Dordrecht, 1994. Kluwer Academic Publishers.

[31] Edgar Soulie. User's experience with Fortran compilers for least squares problems. In Andreas Griewank and George F. Corliss, editors, *Automatic Differentiation of Algorithms: Theory, Implementation, and Application*, pages 297–306. SIAM, Philadelphia, 1991.

[32] Karl Svozil. *Randomness and Undecidability in Physics*. World Scientific, Singapore, 1993.

[33] Olivier Talagrand. The use of adjoint equations in numerical modelling of the atmospheric circulation. In Andreas Griewank and George Corliss, editors, *Automatic Differentiation of Algorithms*, pages 169–180, Philadelphia, 1991. SIAM.

[34] Jean-Noël Thépaut, Drasko Vasiljevic, Philippe Courtier, and Jean Pailleux. Variational assimilation of conventional meteorological observations with a multilevel primitive-equation model. *Q. J. R. Meteorol. Soc.*, 119:153–186, 1993.

[35] Z. Wang, I. M. Navon, X. Zou, and K. J. Ingles. 4-D variational data assimilation with a global multilevel primitive equation model. In *Preprints, Proc. Int. Symp. on Assimilation of Observations in Meteorology and Oceanography, Tokyo, Japan, March 1995*. World Meteorological Organization.

[36] Zhi Wang, Kelvin Droegemeier, Ming Xue, and Seon Ki Park. Sensitivity analysis of a 3-D compressible storm-scale model to input parameters. In *Preprints, Proc. Int. Symp. on Assimilation of Observations in Meteorology and Oceanography, Tokyo, Japan, March 1995*. World Meteorological Organization.

Iterative Domain Decomposition Algorithms: Theory and Applications

Y. Cai and I.M. Navon
Department of Mathematics and
Supercomputer Computations Research Institute
Florida State University
Tallahassee FL 32306-4052 U.S.A.

ABSTRACT. Iterative domain decomposition (DD) methods are a class of "divide and conquer and combine" PDE solving algorithms well suited for implementation on multi-processor architectures. The theory and algorithm of a modified interface matrix (MIM) approach are presented in this contribution as a non-overlapping iterative DD strategy. It is shown in this paper that this novel MIM approach is much more efficient than the well-known Schur DD method for many physical problems where no fast subdomain solvers are available. The resulting numerical algorithm is applied to the efficient parallel finite element solution of the shallow water equations.

1. Introduction and motivation

Recently there has been an increase in research activities in the area of DD methods and parallel computing due to the advent and growth of various parallel processing architectures. The idea of domain decomposition goes back to Schwarz [43] and the work of structural engineers in the early sixties [39]. However, it was only during the last ten years that researchers did develop and extend these ideas for use with multi-processor architectures.

The primary motivation for developing and extending this "old" technique is, no doubt, to take advantage of available modern powerful parallel computers. In fact, DD algorithms were found to be superior to either function-space or operator decomposition from the perspective of parallel computing [23]. There are, however, additional reasons for this revival of interest. Some of these, among others, are listed below:

1. DD techniques offer a possible way to solve PDEs defined on irregular domains with the finite difference or spectral method [26; 41].
2. DD techniques allow us to use different numerical schemes, orders of approximation and resolutions for different subdomains (see, for example, [27]). Thus they offer opportunities to combine the advantages of finite element, spectral and multigrid methods for devising more efficient and accurate algorithms applicable to multi-processor architectures (see [45] and references therein).

F.-X. Le Dimet (ed.), High Performance Computing in the Geosciences, 81–105.

3. The technique also provides us with possible means for isolating regions which require special treatment. Thus the computational effort can be focused upon regions where large gradients, boundary layers, shocks or even singularities occur [15; 28].
4. The technique may be applied to some physical problems which require different mathematical models for different subdomains, for instance, in fluid dynamics, using a viscous model near the boundary and an inviscid model in the far field [12; 16]. Interested readers are also referred to [40] and references therein.

There are two general approaches for DD methods, namely overlapping and non-overlapping, characterized by the way the subdomains are constructed. For the non-overlapping approach, an efficient way to handle the coupling between artificially divided subdomains was first proposed by Dryja in 1982 [13], a work considered to be seminal, in the context of the preconditioned conjugate gradient (PCG) linear symmetric iterative solver. In essence, this is a divide-and-feedback process which continues until a prescribed convergence criterion on the interfaces is attained.

For this Schur DD method, subdomain problems are trivially decoupled after one obtains the boundary conditions, by solving the interface Schur complement matrix linear system, on the artificially introduced interfaces between subdomains. Since the feasibility of this process is based on the property that only matrix-vector products are required in the PCG algorithm for solving a symmetric linear system, the idea is readily extendible to the non-symmetric case.

The iterative solution of the interface linear system, however, usually involves repeated exact subdomain solves which are not cheaply implementable for non-separated elliptic operators or for other more complicated cases such as, for instance, the shallow water equations. To reduce the number of subdomain solves, a lot of research has been carried out in order to find good interface preconditioners in two or three dimensional domains. In this aspect, we mention, among many other papers, [3; 7; 9; 11; 13; 14; 17; 29].

To improve the efficiency of the parallel solution of partial differential equations for which no fast subdomain solvers are available, one may use a DDP (DD preconditioner) approach (also called full matrix DD method in [24]) which was advocated in [4]. The basic idea consists of the construction of a preconditioner in a domain decomposed way such that approximate solutions in the subdomains and on the interfaces can be simultaneously updated at the cost of only inexact subdomain solves. In [25], this approach was applied to some flow problems. Three types of domain decomposed preconditioners have been used to accelerate the convergence of three competitive iterative algorithms for solving non-symmetric linear systems in the context of shallow water flow modeling [5; 6].

In this contribution, we are looking into another possible approach which we call the modified interface matrix (MIM) approach. Like the aforementioned DDP approach, MIM algorithm abandons the idea of Schur DD method that decoupled subdomain problems are independently solved *only after* the solution on the interfaces is obtained. The fundamental idea is to successively improve the subdomain and interface approximate solutions with iterative improvements of the initial solutions in both the subdomains and interfaces being made. Thus, the disadvantage arising from the absence of fast subdomain solvers is mitigated. A detailed comparison between DDP and MIM approaches for the problem at hand will be presented elsewhere in a separate research work.

The current paper is organized in the following way. In section 2, we give formulas corresponding to the substructuring formulation of our problem and briefly describe the Schur DD method. The theory and proposed algorithms for a MIM approach of non-overlapping DD method are presented in section 3. The modified interface matrix is constructed by using a Neumann series expansion and the MILU (modified incomplete LU) factorization in each subdomain. More efficient constructions of the modified interface matrix are currently under further study. Carefully selected numerical results on a single processor and parallelization results on the CRAY Y-MP/432 vector parallel supercomputer are reported and discussed in section 4. Finally, in section 5, the results obtained are summarized and conclusions are drawn.

2. Substructuring and Schur DD method

We only consider solving time dependent PDEs with implicit time discretization. Explicit methods are algebraically equivalent to a matrix-vector product problem and thus readily parallelizable [23]. However, implicit schemes involve matrix inversions and, consequently, are highly sequential. As a standard practice in parallel computing [38], a renumbering or reordering strategy may be used for restructuring the sequence of calculations in such a way as to reveal the parallel structure of the problem.

Since we are interested in implementing our algorithms on parallel computers with powerful vector processors, the original domain is divided into n parallel strips along the west-east directions in order to yield longer vectors [31]. It is clear that cross points are eliminated from our consideration.

If we identify two classes of variables, namely the internal variables relevant to nodes within subdomains and interface variables associated with nodes belonging to two or more subdomains, and number the grid nodes such that internal variables appear before the interface ones (substructure numbering), then the matrix to be inverted at each time step assumes the following block-bordered structure:

$$A = \begin{bmatrix} A_{dd} & A_{ds} \\ A_{sd} & A_{ss} \end{bmatrix} \tag{1}$$

where

$$A_{dd} = \text{diag}[A_{11}, A_{22}, \ldots, A_{nn}] \tag{2}$$

is a block diagonal matrix with each block A_{ii}, for $i = 1, 2, \ldots, n$, being the discrete analog of the restriction of the original differential operator on each subdomain. A_{ds} and A_{sd} are block column and row vectors connecting subdomains to interfaces and vice versa. They assume the following forms

$$A_{ds} = [A_{1s}, A_{2s}, \ldots, A_{ns}]^t \tag{3}$$

$$A_{sd} = [A_{s1}, A_{s2}, \ldots, A_{sn}]. \tag{4}$$

A_{ss} corresponds to the discretization of the original differential operator restricted to the interfaces. Since there are $n-1$ numbers of internal boundaries for the n subdomains case, A_{ss} assumes the following block diagonal form, where each block is associated with one of the interfaces

$$A_{ss} = \mathrm{diag}[T_{11}, T_{22}, \ldots, T_{n-1,n-1}] \tag{5}$$

where the non-zero entries of T_{ii}, $i = 1, 2, \ldots, n-1$, amount to the following cyclic tridiagonal structure induced by the presence of periodic boundary conditions imposed on our physical problem (to be detailed in section 4)

$$T_{ii} = \begin{bmatrix} b_1^{(i)} & c_1^{(i)} & & & a_1^{(i)} \\ a_2^{(i)} & b_2^{(i)} & c_2^{(i)} & & \\ & \ddots & \ddots & \ddots & \\ & & a_{n-1}^{(i)} & b_{n-1}^{(i)} & c_{n-1}^{(i)} \\ c_n^{(i)} & & & a_n^{(i)} & b_n^{(i)} \end{bmatrix} \tag{6}$$

By plotting the surface formed by the entries of a matrix as a function of its indices, the typical block-bordered structure of the matrix A given by (1) corresponding to a substructure numbering of the nodes is generated for a four subdomain case (see Figure 1), where the following modifications are made to the matrix A:

$$a_{ij} = \begin{cases} 1 & \text{if } a_{ij} \neq 0 \\ 0 & \text{if } a_{ij} = 0 \end{cases} \tag{7}$$

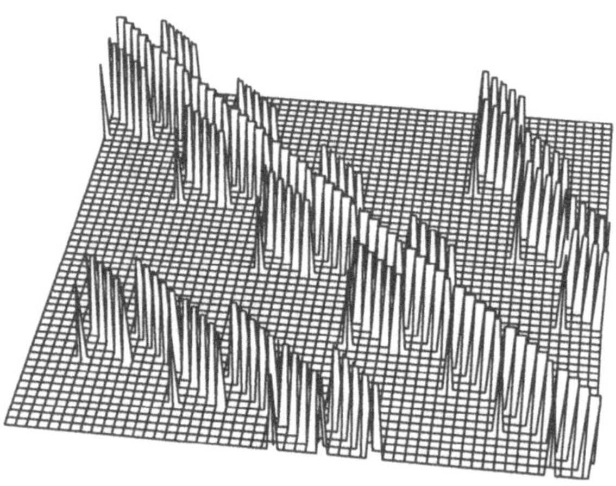

Fig. 1. The typical block-bordered matrix structure corresponding to a substructure numbering of the nodes for a four-subdomain domain decomposition.

The numerical solution of $Ax = f$ is equivalent to solving the following

$$Cx_s = g \qquad \text{on } \Gamma \tag{8}$$

$$A_{ii}x_i = f_i - A_{is}x_s \qquad \text{in } \Omega_i \tag{9}$$

where Γ and Ω_i, $i = 1, 2, \ldots, n$, stand for the interfaces and subdomains, respectively and

$$C = A_{ss} - \sum_{i=1}^{n} A_{si}A_{ii}^{-1}A_{is} \tag{10}$$

and

$$g = f_s - \sum_{i=1}^{n} A_{si}A_{ii}^{-1}f_i \tag{11}$$

where C is the well-known Schur complement matrix.

The Schur DD method starts by first determining boundary conditions x_s on the interfaces between artificially divided subdomains by solving (8). Once x_s is known, the subdomain problems (9) trivially decouple and may be solved in parallel. The iterative algorithms used for solving (8) are those in which only matrix-vector products are required. The matrix-vector product Ct is obtained via solutions in subdomains

$$Ct = A_{ss}t + \sum_{i=1}^{n} A_{si}v_i \tag{12}$$

where v_i, for $i = 1, 2, \ldots, n$, are solutions of

$$A_{ii}v_i = -A_{is}t \qquad \text{in subdomains } \Omega_i \tag{13}$$

Since the linear system to be solved is non-symmetric due to the presence of advection terms in the PDEs, we apply the PCGS (preconditioned conjugate gradient squared) algorithm [44], which assumes the following algebraic form ready for computer implementation (for solving the linear system $Cx_s = g$):

$$\begin{gathered}
r = g, \qquad x_s = x_s^0, \qquad r = r - Cx_s \\
\text{Choose } \tilde{r} \text{ such that } (\tilde{r}, r) \neq 0 \\
p = r, \qquad u = r \\
t = G^{-1}p \\
\text{p1} = Ct, \qquad \delta 0 = (\tilde{r}, \text{p1}), \qquad \delta 1 = (\tilde{r}, r) \\
\alpha = \delta 1/\delta 0, \qquad q = u - \alpha \text{p1}, \qquad u = u + q \\
t = G^{-1}u, \qquad x_s = x_s + \alpha t
\end{gathered} \tag{14}$$

If the convergence criterion is met then stop, otherwise continue

$$\begin{gathered}
\text{p1} = Ct, \qquad \delta 0 = (\tilde{r}, r), \qquad r = r - \alpha \text{p1} \\
\delta 1 = (\tilde{r}, r), \qquad \beta = \delta 1/\delta 0
\end{gathered}$$

$$u = r + \beta q, \qquad p = u + \beta(q + \beta p)$$

Goto (14)

where G is an appropriate preconditioner.

We mention here that there are other suitable, sophisticated and competitive iterative algorithms for solving non-symmetric linear systems. For instance, GMRES (generalized minimum residual) [42] or the recently proposed Bi-CGSTAB (a fast and smoothly convergent variant of bi-conjugate gradient method) [47]. These three iterative algorithms, namely, GMRES, CGS and Bi-CGSTAB, with block post-preconditioning have recently been applied to the shallow water finite element simulation problem (see [5; 6] for details).

The Schur DD method may be heuristically described as a divide-and-feedback (or "divide and conquer and combine") process. The problem defined on the original domain is divided into smaller problems and solved in parallel within the subdomains. The information gathered from the solution of each of these subdomain problems is then fed back to the interface iterative solver. If the convergence criterion is not met on the interfaces, the domain is decomposed again and the subdomain problems solved. This divide-and-feedback process continues until the convergence criterion on the interfaces is fulfilled.

This divide-and-feedback process indicates that the efficiency of this approach relies on the number of iterations required for obtaining convergence on the interfaces as well as on the computational cost for each subdomain solver. Preconditioning techniques may be used to reduce the number of iterations on the interfaces. In the case where fast solvers do not exist, either direct solvers based on LU factorizations or iterative methods may be applied in the subdomains.

For the iterative subdomain solver, preconditioning based on either ILU [30] or MILU [2; 20; 21] factorizations may be applied. The MILU factorization preconditioning was proposed so that the spectral condition number of the matrix $G^{-1}A$ satisfies the following asymptotic relation:

$$\kappa(G^{-1}A) \sim O(h^{-1}), \quad as\ h \to 0 \tag{15}$$

where h is the mesh size, instead of $O(h^{-2})$ as for ILU. Here $G = LU = A + D + R$, R is the so-called defect or error matrix whose rowsum is zero for each row of R. D is a diagonal matrix containing some preconditioning parameters.

Specifically, for a class of M-matrices or a weakly diagonally dominant symmetric L-matrices, it was shown that the MILU factorization could be constructed such that (15) holds. For some other more general coefficient matrices, as pointed out in [20], the same rate of convergence was observed.

It is readily observed that, at each time step, the incomplete factorization has to be carried out only once for each subdomain during the entire divide-and-feedback process. The work required for the preconditioning of each subdomain consists mainly of the forward and back substitutions which may be computed in parallel.

For the interfaces, a class of relatively robust preconditioners is the so-called interface probe preconditioners (also called modified Schur complement preconditioners in [24]) which were first proposed in [10] and discussed in some detail in

[25]. The basic idea involved in the interface probe construction is to capture the strong dependence between neighboring grid nodes and ignore the weak dependencies between nodes which are mesh-size distance apart on the interfaces. Reference [9] contains the most recent and comprehensive review and update of the interface probe preconditioning techniques.

Following the notations used in [8], IP(0) and IP(1) are the most commonly used among many choices available within this class. However, we will make some modifications to IP(0) and construct our interface preconditioner by retaining the cyclic tridiagonal structure of each T_{ii} in A_{ss} and then replacing each entry in the main diagonal of A_{ss} by the corresponding row-sum of the Schur complement matrix C. We denote this interface preconditioner by MIP(0). Our experiences indicate that MIP(0) is better than IP(0).

$3n$ or n subdomain solves are required for constructing the interface preconditioner using IP(1) or MIP(0), respectively, compared to $n_s n$ subdomain solves required for constructing the Schur complement matrix. Numerical experience with MIP(0) and IP(1) for solving the shallow water equations is reported in section 4. These two types of interface preconditioners preserve the block diagonal structure of the matrix A_{ss}. We denote them in a unified way by

$$G = diag[\tilde{T}_{11}, \tilde{T}_{22}, \ldots, \tilde{T}_{n-1,n-1}] \tag{16}$$

Then, the preconditioning linear system $Gu = v$ may be split up into $n-1$ smaller systems $\tilde{T}_{ii}u_i = v_i$, $i = 1, 2, \ldots, n-1$, which can be solved in parallel. Since each of these $\tilde{T}_{ii}$ is cyclic tridiagonal, we use the so-called (see [46]) Ahlberg-Nielson-Walsh algorithm [1], which is an extension of the well-known double-sweep algorithm of Thomas [48]. For an efficient numerical algorithm for solving cyclic pentadiagonal linear systems, we recommend [36].

To facilitate the comparison of various numerical experiments to be reported in section 4, we use the following definitions:

1. SDD I — MILU-PCGS in the subdomains and no interface preconditioning;
2. SDD II — MILU-PCGS in the subdomains and interface preconditioning by IP(1).
3. SDD III — MILU-PCGS in the subdomains and interface preconditioning by MIP(0).

3. The MIM approach

For the Schur DD approach, the subdomain problems (9) are solved only after the solution x_s on the interfaces is obtained. We propose, in this section, a new approach to handle the coupling between different subdomains. It differs from the Schur DD method in that the approximations to the solutions on the interfaces and in the subdomains are successively improved.

Theorem 1 *Sequences of approximation $x_i^{(k)}$, for $i = 1, 2, \ldots, n$, and $x_s^{(k)}$ produced by*

$$For\ k = 0, 1, 2, \ldots$$

$$A_{ii}x_i^{(k+1)} = f_i - A_{is}x_s^{(k)} \qquad for\ i = 1, 2, \ldots, n \tag{17}$$

$$A_{ss}x_s^{(k+1)} = f_s - \sum_{i=1}^{n} A_{si}x_i^{(k+1)} \tag{18}$$

converge to x_i^*, for $i = 1, 2, \ldots, n$, and x_s^* for an arbitrary initial solution $x_s^{(0)}$ on the interfaces if and only if the spectral radius of the matrix $I_{ss} - A_{ss}^{-1}C$ satisfies

$$\rho(I_{ss} - A_{ss}^{-1}C) < 1 \tag{19}$$

where I_{ss} is an identity matrix of size $n_s \times n_s$, x_s^* and x_i^*, for $i = 1, 2, \ldots, n$, are the solutions of (8) and (9).

PROOF. Define an error vector $e_s^{(k)} = x_s^{(k)} - x_s^*$ for the interface approximation. Since $x_i^{(k+1)} = A_{ii}^{-1}(f_i - A_{is}x_s^{(k)})$, for $i = 1, 2, \ldots, n$, it is straightforward to show that $A_{ss}x_s^{(k+1)} = g + (A_{ss} - C)x_s^{(k)}$, where g is defined by (11). It follows $e_s^{(k+1)} = (I_{ss} - A_{ss}^{-1}C)e_s^{(k)}$. Hence we have $x_s^{(k)} \to x_s^*$, as $k \to \infty$ if and only if (19) holds. Now let $k \to \infty$ in (17), it follows from (9) that $x_i^{(k)} \to x_i^*$, as $k \to \infty$, for $i = 1, 2, \ldots, n$.

By theorem 1, we obtain the following iterative procedure for solving the linear system $Ax = f$. The iteration starts with an arbitrary initial solution $x_s^{(0)}$ on the interfaces $\Gamma = \Gamma_1 \cup \Gamma_2 \cup \cdots \cup \Gamma_{n-1}$, solves the subdomain problems (17) in parallel, then updates the approximation on the interfaces by solving (18). In this way we solve the subdomain problems and the interface problem successively until the convergence criterion on the interfaces is met.

Compared with the Schur DD method, the algorithm based on theorem 1 is quite straightforward and the condition (19) is satisfied in our case for various mesh resolutions. Specifically, our numerical experiments (see section 4) indicate that for various mesh resolutions of the discretized domain, the spectral radii $\rho(I_{ss} - A_{ss}^{-1}C)$ are about 0.13. Hence the asymptotic rate of convergence $-\ln\rho$, (see, for example, ([22])) is about 2.0402. Thus, in order to reduce the norm of the initial error vector on the interfaces by a factor of, say, 10^{-6}, roughly seven iterations will be required.

We now proceed to modify this iterative procedure. Since each iteration using (17) and (18) requires the solutions of all the subdomain problems once, it is important to reduce the number of iterations. It is well known that the smaller the spectral radius, the faster the rate of convergence. In order to reduce the spectral radius, we construct a matrix K_{ss} such that $A_{ss}+K_{ss}$ is a good approximation of the Schur complement matrix C in the sense that the spectral radius $\rho[(A_{ss}+K_{ss})^{-1}C]$ is close to 1. The matrix $A_{ss} + K_{ss}$ is referred to as the modified interface matrix.

Now assume that the matrix K_{ss} has been chosen, the following result may be proved in a similar way

Theorem 2 *Sequences of approximation $x_i^{(k)}$, for $i = 1, 2, \ldots, n$, and $x_s^{(k)}$ obtained by*

$$For\ k = 0, 1, 2, \ldots$$

$$A_{ii}x_i^{(k+1)} = f_i - A_{is}x_s^{(k)} \qquad for\ i = 1, 2, \ldots, n \tag{20}$$

$$(A_{ss}+K_{ss})\Delta x_s^{(k)} = f_s - A_{ss}x_s^{(k)} - \sum_{i=1}^{n} A_{si}x_i^{(k+1)} \tag{21}$$

converge to x_i^*, for $i = 1, 2, \ldots, n$, and x_s^* for an arbitrary initial solution $x_s^{(0)}$ on the interfaces if and only if the spectral radius of the matrix $I_{ss} - (A_{ss}+K_{ss})^{-1}C$ satisfies

$$\rho(I_{ss} - (A_{ss}+K_{ss})^{-1}C) < 1 \tag{22}$$

where $\Delta x_s^{(k)} = x_s^{(k+1)} - x_s^{(k)}$, I_{ss} is an identity matrix of size $n_s \times n_s$, x_s^* and x_i^*, for $i = 1, 2, \ldots, n$, are the solutions of (8) and (9). Moreover, the following norm relations hold

$$\|(A_{ss}+K_{ss})\Delta x_s^{(k)}\| = \|g - Cx_s^{(k)}\| \tag{23}$$

where C is the Schur complement matrix and g is the corresponding right hand side, defined by (11).

According to theorem 2, the iteration starts with an arbitrary initial solution $x_s^{(0)}$ on the interfaces and we successively solve the subdomain problems (20) in parallel and the linear system (21) on the interfaces until the norm given by (23) is sufficiently reduced.

It is interesting to make an analogy here and think of each of the equations in (20) as the governing equation for the displacement distribution within a substructure subjected to a load f_i originally acting upon it and the boundary interaction forces $-A_{is}x_s^{(k)}$ due to interface connections between the substructures.

By taking $x_s^{(0)}$ to be some initial solution on the interfaces that differs from the true solution x_s^*, we are actually imposing some constraints in addition to the original constraints (the boundary conditions) to the structure and thus making it stiffer. However, as guaranteed by theorem 2 the extra constraints introduced due to the incorrect initial solution will be continually relaxed by solving the linear system (21) repeatedly.

In the sequel our discussion will be related to the CGS non-symmetric iterative method for each of the linear systems (20) in the subdomains, preconditioned by the MILU factorization, and for system (21) on the interfaces accelerated by a preconditioner constructed in a similar way to that of MIP(0).

We now investigate the construction of the matrix K_{ss} such that the modified interface matrix $A_{ss}+K_{ss}$ constitutes a good approximation of the Schur complement matrix C. Since we are mainly interested in non-symmetric iterative methods for which only matrix-vector products are required for the solution of the system (21), it is more efficient to consider the algorithm of computing $K_{ss}w_s$ for a given vector w_s rather than first forming the matrix and then the product.

As a first approach, let us consider a splitting of each of the subdomain matrices A_{ii}

$$A_{ii} = P_{ii} - Q_{ii} = P_{ii}(I_{ii} - R_{ii}) \tag{24}$$

where P_{ii} is a nonsingular matrix and I_{ii} is the identity matrix of size $n_i \times n_i$. Assuming that the spectral radius of $R_{ii} = I_{ii} - P_{ii}^{-1}A_{ii}$ is less than 1, we may obtain the following Neumann series expansion

$$A_{ii}^{-1} = [\sum_{k=0}^{\infty}(I_{ii} - P_{ii}^{-1}A_{ii})^k]P_{ii}^{-1} \tag{25}$$

$C - A_{ss}$ can be approximated by using only a finite truncated expansion, i.e.

$$C - A_{ss} \approx K_{ss} = -\sum_{i=1}^{n} A_{si}X_{is}^{(m+1)} \tag{26}$$

where

$$X_{is}^{(m+1)} = [\sum_{k=0}^{m}(I_{ii} - P_{ii}^{-1}A_{ii})^k]P_{ii}^{-1}A_{is} \tag{27}$$

The following algorithm can be shown to be valid for computing $K_{ss}w_s$

$$K_{ss}w_s = -\sum_{i=1}^{n} A_{si}v_i^{(m+1)} \tag{28}$$

where the vector $v_i^{(m+1)}$ can be obtained by the following iterative procedure: for $k = 0, 1, \ldots, m$, starting with $v_i^{(0)} = 0$

$$P_{ii}(v_i^{(k+1)} - v_i^{(k)}) = -A_{ii}v_i^{(k)} + u_i \tag{29}$$

for $i = 1, 2, \ldots, n$ and $u_i = A_{is}w_s$. Notice that (29) can be implemented and $A_{si}v_i^{(m+1)}$ formed, for $i = 1, \ldots, n$, completely in parallel.

Another approach is to use MILU preconditioners in the subdomains $A_{ii} \approx L_iU_i$ to construct the matrix K_{ss}. Specifically, we take

$$K_{ss} = -\sum_{i=1}^{n} A_{si}(L_iU_i)^{-1}A_{is} \tag{30}$$

Here the matrix-vector product may be evaluated as follows:

$$K_{ss}w_s = -\sum_{i=1}^{n} A_{si}v_i \tag{31}$$

where v_i can be determined by

$$L_i\hat{v}_i = A_{is}w_s \tag{32}$$

and

$$U_iv_i = \hat{v}_i \tag{33}$$

for $i = 1, 2, \ldots, n$. The solutions of (32) and (33) can be carried out and $A_{si}v_i$ formed in parallel.

Numerical results indicate that the spectral radii $\rho(I_{ss} - (A_{ss} + K_{ss})^{-1}C)$ for various resolutions are around .04 by retaining just the first term in the Neumann series and 0.002 by using the MILU factorization. Thus, roughly only four and two iterations, respectively, will be required in order to reduce the norm of the initial error vector on the interfaces by a factor of 10^{-6},

Prior to presenting an algorithm of the modified interface matrix approach, we discuss here how to choose initial guesses for the solution of (20) as well as that of (21) and how to precondition (21).

Rather than making a guess of the initial vector for the PCGS solver to start with, we take $x_i^{(k)}$, for $i = 1, 2, \ldots, n$ as the initial guess for the solution of each of the linear systems in (20), for $k = 1, 2, \ldots$, as well as the zero vector as the initial guess for the modified interface matrix linear system (20), for $k = 0, 1, \ldots$.

It is clear from theorem 2 that, as k increases, $x_i^{(k+1)} - x_i^{(k)} \to 0$, for $i = 1, 2, \ldots, n$, and $\Delta x_s^{(k)} \to 0$ on the interfaces. Consequently, the initial vectors selected this way become better approximations to the subdomain solutions $x_i^{(k+1)}$, for $i = 1, 2, \ldots, n$, as well as to the solution on the interfaces $\Delta x_s^{(k)}$, as the iterative procedure defined by (20) and (21) proceeds, and thus fewer iterations are required for the iterative solutions in the subdomains (20) and on the interfaces (21). As a result, the computational cost of the MIM approach decreases as k increases. The reduced cost for the subdomains mitigates the disadvantage of unavailability of fast subdomain solvers.

Here we use, for any fixed k, essentially the same preconditioner as MIP(0) previously used for preconditioning (8), with the Schur complement matrix replaced by the matrix $A_{ss} + K_{ss}$. Only a single probing vector of the form $v = \{1, 1, 1, 1, 1, 1, \ldots\}^t$ is required for this construction. The product $K_{ss}v$ is evaluated by using either (28) or (31). If (31) is used, it is easy to see that just one inexact solve is required for the construction of this preconditioner in each subdomain.

Based on the above discussion, we present in the following an algorithm for the modified interface matrix approach, where steps 1 to 3 are for set-up purposes. The iterative procedure is described by steps 4 through 6.

Algorithm MIMA (Modified Interface Matrix Approach)

Step 1. Carry out the MILU factorizations of A_{ii}, $A_{ii} \approx L_i U_i$, for $i = 1, 2, \ldots, n$, in parallel.

Step 2. Construct the preconditioner for (21) similar to MIP(0). A large part of this calculation may be carried out in parallel by using either (28) or (31).

Step 3. Set $k = 0$. Specify $x_s^{(0)}$ on the interfaces and solve subdomain problems (20) in parallel by using the MILU PCGS solver with a suitable initial solution.

Step 4. Solve (21) by the PCGS solver with the preconditioner constructed in step 2, and an initial solution taken to be the zero vector on the interfaces. The preconditioning system may be solved in parallel and the matrix-vector product $K_{ss}w_s$ may be computed by either (28) or (31) mostly in parallel, where w_s varies between iterations.

Step 5. Test for convergence on the interfaces (see (23). If the convergence criterion is met, go to step 6 and stop; Otherwise, go to step 6 and then go back to step 4.

Step 6. Set $k \leftarrow k+1$. Solve subdomain problems (20) in parallel by using the MILU PCGS solver with initial solutions $x_i^{(k)}$, for $i = 1, 2, \ldots, n$.

For numerical comparisons carried out in section 4, algorithm MIMA with $K_{ss}w_s$ computed by either (28) or (31) will be referred to as MIMA I or MIMA II, respectively.

4. A physical model and numerical results

The shallow water equations model, in its primitive formulation, assumes the following form

$$\frac{\partial U}{\partial t} + \begin{bmatrix} u & \varphi & 0 \\ 1 & u & 0 \\ 0 & 0 & u \end{bmatrix} \frac{\partial U}{\partial x} + \begin{bmatrix} v & 0 & \varphi \\ 0 & v & 0 \\ 1 & 0 & v \end{bmatrix} \frac{\partial U}{\partial y} + \begin{bmatrix} 0 & 0 & 0 \\ 0 & 0 & -f \\ 0 & f & 0 \end{bmatrix} U = 0 \tag{34}$$

or

$$\frac{\partial U}{\partial t} + A\frac{\partial U}{\partial x} + B\frac{\partial U}{\partial y} + CU = 0 \tag{35}$$

where $U = (\varphi, u, v)^T$. $\varphi = gh$ is the geopotential of the fluid, g is the acceleration of gravity and h is the height of the free surface. f is the Coriolis parameter given by a β plane approximation, while u and v are the horizontal velocity components in x and y directions, respectively.

The model assumes that the fluid is homogeneous, inviscid, barotropic and incompressible. These pure convection equations are defined on a limited-area rectangular domain which corresponds to a channel on the rotating earth. The southern and northern boundaries are assumed to be rigid, i.e., $v = 0$ and the flow is assumed periodic in the west-east direction. This is a standard non-linear shallow water equations model which was used to test various finite difference schemes in [18].

The usual Galerkin finite element procedure was used for the spatial discretization with linear triangular elements. An extrapolated Crank-Nicolson scheme was employed for the time discretization (see [33; 37; 34; 35] for details). After collecting and rearranging terms, we obtain the following three linear systems which need to be inverted at each time step

$$A^n \Delta\varphi^n = f_\varphi^n, \quad B^n \Delta u^n = f_u^n, \quad C^n \Delta v^n = f_v^n \tag{36}$$

where A^n, B^n and C^n are nonsymmetric matrices due to the presence of advective terms and $\Delta\varphi^n = \varphi^{n+1} - \varphi^n$, $\Delta u^n = u^{n+1} - u^n$, $\Delta v^n = v^{n+1} - v^n$. Three right hand side terms in (36) are functions of nodal unknowns at the previous steps.

We carried out numerical experiments on the Cray Y–MP/432 which has a machine accuracy around 0.8×10^{-14} for a single precision. For our numerical experiments, we used the following constants whose numerical values are summarized below

$$L = 6000\ km, \qquad D = 4400\ km$$
$$g = 10\ m/s, \qquad H_0 = 2000\ m$$
$$H_1 = 220\ m, \qquad H_2 = 133\ m$$
$$\hat{f} = 10^{-4}\ s^{-1}, \qquad \beta = 1.5 \times 10^{-11}\ s^{-1}m^{-1}$$

where L and D are the length and width of the channel; g is the acceleration of gravity; $\hat{f}$ and β are used to determine the the Coriolis parameter $f = \hat{f}+\beta(y-D/2)$ and finally three constants H_0, H_1 and H_2 are used to analytically define the initial geopotential and velocity fields by the following geostrophic relationship

$$\varphi = gh, \qquad u = -(g/f)\frac{\partial h}{\partial y}, \qquad v = (g/f)\frac{\partial h}{\partial x} \tag{37}$$

where the initial height field is given by

$$\begin{aligned} h(x,y) &= H_0 + H_1 \tanh[9(D/2-y)/2D] \\ &\quad + H_2 \mathrm{sech}^2[9(D/2-y)/2D]\sin(2\pi x/L). \end{aligned} \tag{38}$$

The geopotential field distribution in the present problem has the order of magnitude $10^4\ m^2/s^2$. We scale our problem by introducing the following dimensionless variables:

$$x' = x/L, \qquad y' = y/L, \tag{39}$$
$$t' = t\sqrt{\varphi_0}/L, \qquad \varphi' = \varphi/\varphi_0, \tag{40}$$
$$u' = u/\sqrt{\varphi_0}, \qquad v' = v/\sqrt{\varphi_0}, \tag{41}$$
$$h' = h/L, \qquad H_0' = H_0/L, \tag{42}$$
$$H_1' = H_1/L, \qquad H_2' = H_2/L, \tag{43}$$
$$g' = gL/\varphi_0, \qquad f' = fL/\sqrt{\varphi_0}. \tag{44}$$

The pre–selected reference geopotential has been chosen to be $\varphi_0 = 10^4\ m^2/s^2$ so that the non-dimensionalized geopotential φ' as defined by (40) has the order of magnitude of $O(1)$. Under this choice of reference geopotential the dimensionless constants become

$$L' = 1 \qquad D' = 0.733333$$
$$g' = 6000 \qquad H_0' = 0.333333 \times 10^{-3}$$
$$H_1' = 0.366666 \times 10^{-4} \qquad H_2' = 0.221666 \times 10^{-4}$$
$$\hat{f}' = 6 \qquad \beta' = 5.4$$

The initial non-dimensionalized geopotential distribution for the 2-D shallow water equations under consideration is shown in Figure 2. The whole domain is divided into four subdomains of equal sizes with three internal boundaries which collectively constitute the interfaces of the domain. In Figure 3, a typical finite-element domain decomposition solution is presented for the non-dimensionalized geopotential distribution of the shallow water equations model on a limited-area domain at the end of five hours of numerical integration. A mesh resolution 45 by 39 was used to produce the solution. There are 405 nodes in each of the four subdomains and 135 nodes on the interfaces. This run requires a total of 1755 nodes and 3420 elements in the whole domain. The time step is half an hour or .03 in non-dimensionalized time.

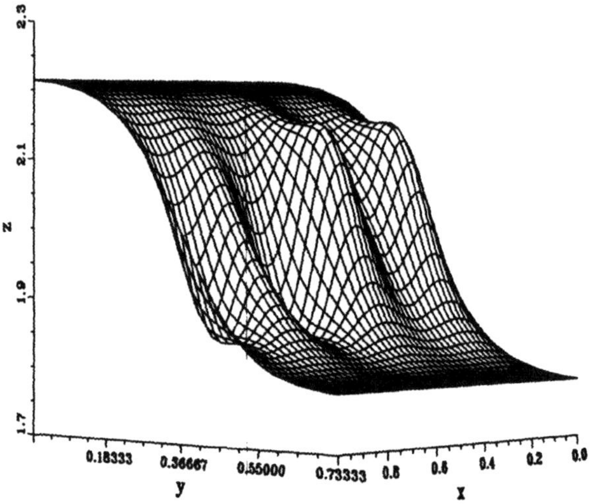

Fig. 2. The initial non-dimensionalized geopotential field.

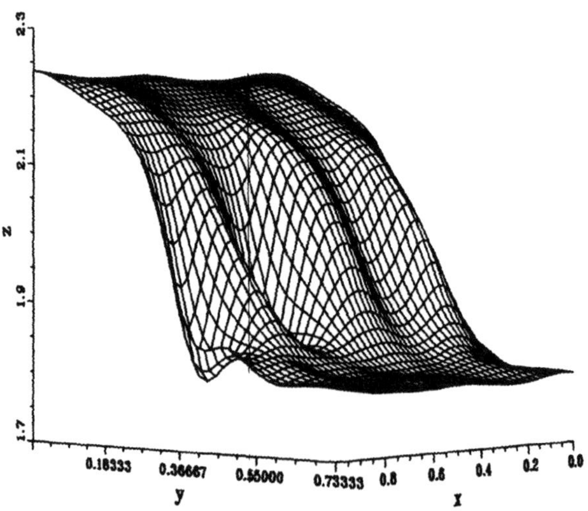

Fig. 3. The non-dimensionalized geopotential field after five hours of numerical integration using the finite element model with four-subdomain domain decomposition.

It should be pointed out that, although all of the algorithms presented in this paper may be applied to finite element domain decomposition solutions of the shallow water equations and yield similar numerical results, the respective computational costs corresponding to the implementation of different algorithms are vastly different.

In the sequel, we will compare the relative efficiencies of various algorithms. For convenience, the Schur DD approach and the MIM approach have been abbreviated to SDD and MIMA. The 2-norm is used throughout and the stopping criterion for the iterative algorithm is based on the 2-norm of the final residual vector being less than 10^{-10} unless indicated otherwise.

To justify our choosing to apply the interface probe preconditioners to the iterative solution of Schur linear systems, we present, in Figure 4, the surface generated by the entries in the non-dimensionalized geopotential Schur complement matrix as a function of its indices at the end of one hour. The mesh resolution used for this calculation is 31 by 31 grid points. However, similar diagonally-dominant structures were observed for other mesh resolutions.

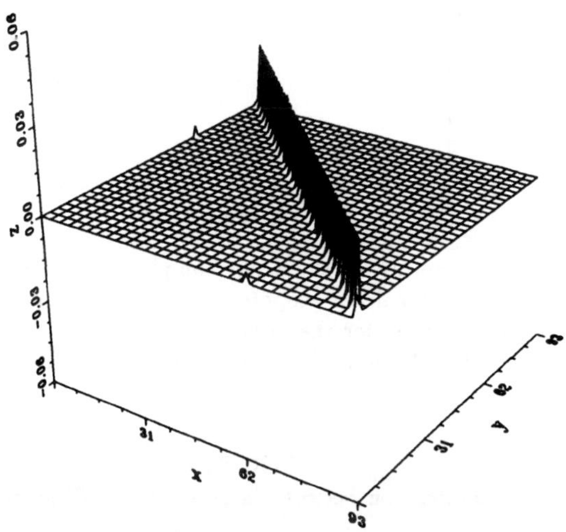

Fig. 4. The surface generated by the non-dimensionalized geopotential Schur complement matrix at the end of one hour. The mesh resolution is 31 by 31 for the original domain. For this choice, there are 93 nodes on the interfaces and 217 nodes in each of the four subdomains.

It was also found that structures of the Schur complement matrices governing the velocity distributions on the interfaces and corresponding to different mesh resolutions are all similar to the tridiagonal structure displayed in Figure 4. This typical structure of the Schur complement matrix was first observed in [17] for the case of the Laplace operator.

We tested the Schur DD method corresponding to various mesh resolutions using SDD II and SDD III algorithms (see the end of section 2). It was found that the

SDD III algorithm produced invariably better results in terms of both the number of iterations required to attain convergence and the corresponding CPU time.

We present, in Figure 5, the log_{10} norms of the residual vectors on the interfaces vs. the number of iterations corresponding to each of these two cases. A mesh resolution of 51 by 51 was used for this calculation. For comparison purposes, the case corresponding to the SDD I algorithm is also included. Both the number of iterations and the CPU time required for each case were recorded in Table I for the solution of the non-dimensionalized geopotential Schur complement linear system at the end of one hour.

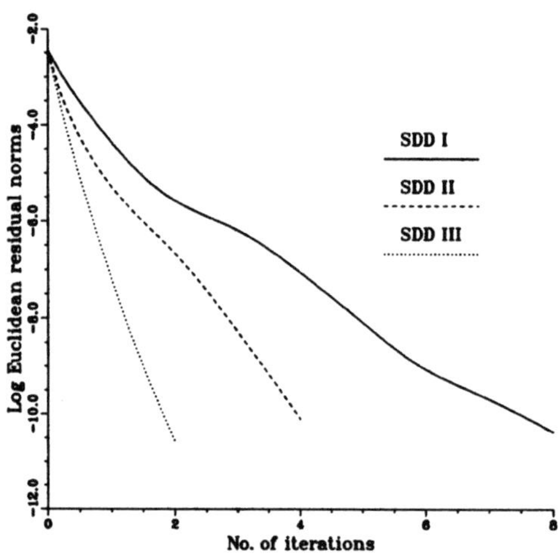

Fig. 5. The log_{10} residual norms for the Schur complement matrix linear system on the interfaces for the non-dimensionalized geopotential matrix system at the end of one hour. The mesh resolution is 51 by 51 for the original domain. For this choice, there are 153 nodes on the interfaces and 612 nodes in each of the four subdomains.

TABLE I
A comparison between interface preconditioners

SDD I	SDD II	SDD III
8 iterations	4 iterations	2 iterations
4.08 seconds	2.12 seconds	1.02 seconds

A typical convergence behavior corresponding to MILU preconditioning is displayed in Figure 6, where the mesh resolution is 51 by 51 and the stopping criterion has been chosen to be that the 2-norm of the final residual vector is less that 10^{-15}. The results, displaying the numbers of iterations and the CPU time required to attain convergence, are presented in Table II for the solution of a typical subdomain.

We turn now our attention to the theory discussed in section 3. The critical factor that affects the successful use of the theory and algorithms presented there

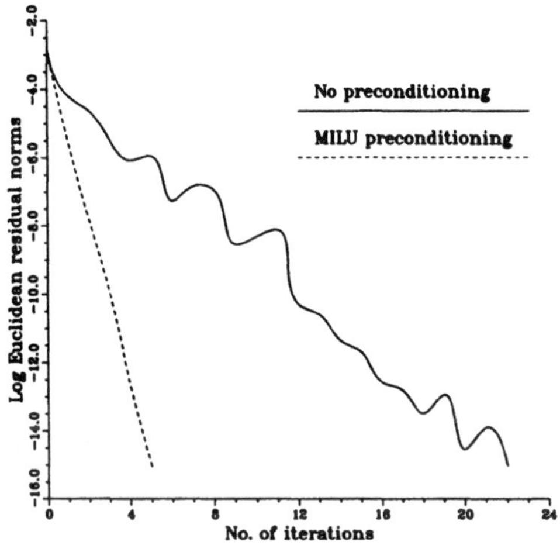

Fig. 6. The log_{10} residual norms in the subdomain for the non-dimensionalized geopotential matrix system at the end of one hour with and without preconditioning. The mesh resolution is 51 by 51 for the original domain and for this choice there are 612 nodes in each of the four subdomains.

TABLE II
MILU preconditioning in a typical subdomain

No preconditioning	MILU preconditioning
22 iterations	5 iterations
0.2207 seconds	0.1480 seconds

is the spectral radii of the matrices $(I_{ss} - A_{ss}^{-1}C)$ and $(I_{ss} - (A_{ss} + K_{ss})^{-1}C)$, respectively. Our numerical experiments with various mesh resolutions indicate that condition (19) is satisfied and that the effect of modifying A_{ss} to $A_{ss} + K_{ss}$ is computationally significant.

We present numerical results obtained using two mesh resolutions, one being 21 by 19, while the other being 41 by 39. For these two cases we obtained that $\rho(I_{ss} - A_{ss}^{-1}C) = 0.1349$ and 0.1340 respectively. For the construction of the matrix K_{ss}, both the MILU factorizations and Neumann series expansions in the subdomains are used. In the case of Neumann series expansion, P_{ii}, for $i = 1, 2, \ldots, n$, has been chosen to be the diagonal part of the matrix A_{ii} (see (24)). The results for $\rho(I_{ss} - (A_{ss} + K_{ss})^{-1}C)$ are summarized in Table III, where the second row in the table corresponds to the spectral radii for the lower resolution, while the third row corresponds to the higher resolution.

We notice that the matrix K_{ss} constructed by MILU factorization in the subdomains yields the smallest spectral radius and therefore a faster convergence rate of the iterative procedure is to be expected. For $m = 3$ case, i.e., by retaining

TABLE III

A comparison between spectral radii by using MILU factorization and Neumann series expansion

MILU	$m = 0$	$m = 1$	$m = 2$
2.127×10^{-3}	4.138×10^{-2}	2.750×10^{-2}	1.867×10^{-2}
1.952×10^{-3}	3.922×10^{-2}	2.571×10^{-2}	1.662×10^{-2}

four terms in the Neumann series expansion, the spectral radii for the lower and higher resolutions are, respectively, 1.508×10^{-2} and 1.199×10^{-2}. They are still not as small as those obtained using MILU factorizations. Moreover, the CPU time consumed by taking four or more terms in the Neumann series generally outweighs the gain obtained by the reduction of the number of iterations.

The modified interface matrix $A_{ss} + K_{ss}$ constructed according to (30) constitutes quite a good approximation to the Schur complement matrix. A plot of the surface formed by the elements in the modified interface matrix as a function of its indices can hardly be distinguished from Figure 4. Instead we present, in Figure 7, the surface generated by the entries in the matrix $C - (A_{ss} + K_{ss})$ corresponding to the non-dimensionalized geopotential matrix at the end of one hour. Similar structures are observed for other mesh resolutions.

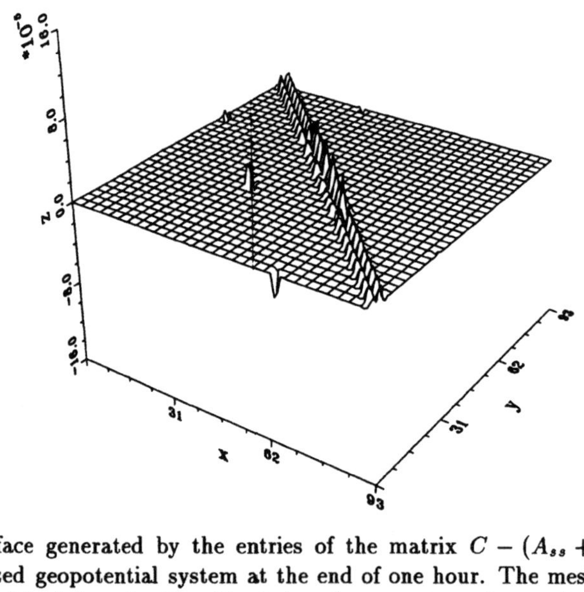

Fig. 7. The surface generated by the entries of the matrix $C - (A_{ss} + K_{ss})$ for the non-dimensionalized geopotential system at the end of one hour. The mesh resolution is 31 by 31 for the original domain. For this choice, there are 93 nodes on the interfaces and 217 nodes in each of the four subdomains.

To compare the performance behavior of the MIMA I and II algorithms, we present, in Figure 8, the log_{10} norms of the residual vectors (see (23) on the interfaces for various cases, where the IMA (Interface Matrix Approach), the approach based on Theorem 1, is also included for comparison. We see that the modified interface matrix formed by the MILU factorization in each subdomain yields the best results. This is to be expected since $\rho(I_{ss} - (A_{ss} + K_{ss})^{-1}C)$ is of order 10^{-3} (see Table III).

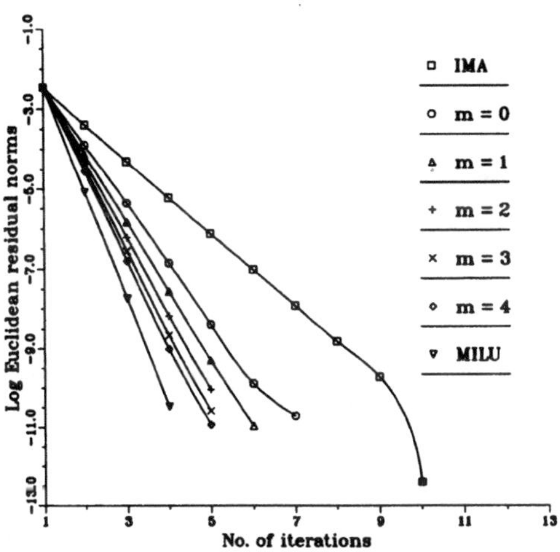

Fig. 8. The log_{10} residual norms for the modified interface matrix system on the interfaces for the non-dimensionalized geopotential matrix system at the end of one hour using the algorithm MIMA. The mesh resolution is 51 by 51. For this choice, there are 153 nodes on the interfaces and 612 nodes in each of the four subdomains.

The number of iterations and the CPU time required for the solution of the geopotential distribution at the end of one hour corresponding to each one of these cases are summarized in Table IV, where IMA is the case when $K_{ss} = 0$; $m = 0, 1, \ldots, 4$ correspond to K_{ss} constructed according to (26) by retaining one, two, ..., five terms in the Neumann series expansion respectively.

TABLE IV
A comparison between algorithm MIMA I and II

IMA	$m=0$	$m=1$	$m=2$	$m=3$	$m=4$	MILU
10	7	6	5	5	5	4
2.435	1.895	1.834	1.729	1.779	1.843	1.607

To see how the initial guesses in the subdomains and on the interfaces improve as the iteration count k in algorithm MIMA increases, we carried out numerical experiments based on the algorithm MIMA II and kept a record of the initial residual norms for each of the four subdomain problems and the modified interface matrix linear system as functions of the number of iterations k (see Figure 9 for details).

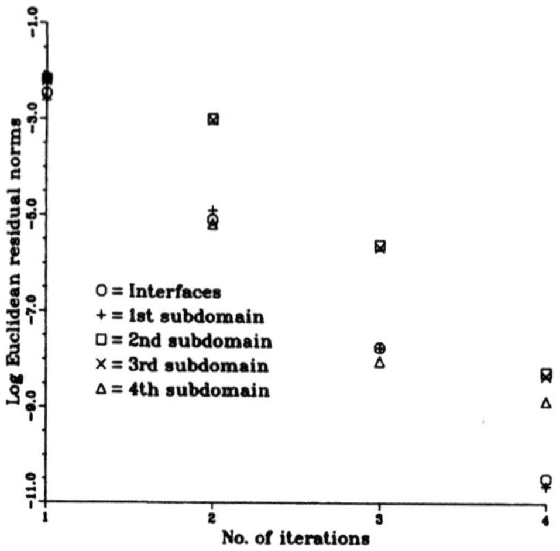

Fig. 9. The history of improved initial guesses in the subdomains and on the interfaces for the non-dimensionalized geopotential matrix system at the end of one hour using the MIMA II algorithm. The mesh resolution is 51 by 51 in the original domain.

Table V is presented here to provide further information. We integrated the shallow water equations model by using algorithm MIMA II for one hour for five different mesh resolutions with and without the improved initial guesses in the subdomains. These two cases are identified by MIMA II(a) and MIMA II(b). For larger problems, the subdomain solvers become more expensive and hence differences of the computational cost between MIMA II(a) and II(b) become more pronounced as confirmed by the results summarized in Table V. For the case in which no improvements are made, the initial guesses are taken to be zero vectors for the subdomain solvers.

TABLE V
The effect of the improved initial guesses

resolutions	CPU time for MIMA II(a)	CPU time for MIMA II(b)
15×15	0.418	0.525
27×27	1.515	1.886
39×39	3.854	4.936
51×51	8.879	11.205
63×63	17.537	23.349

At this point it is natural to compare the SDD III with the MIMA II algorithms by integrating the shallow water equations model for a period of one hour with various mesh resolutions. It becomes evident that, as the mesh resolution increases, the computational costs of the solutions in the four subdomains and on the interfaces will play a prominent role in the computational efficiency of the entire integration process. Thus we expect the MIMA II algorithm to behave much better in terms of CPU time than SDD III algorithm for higher mesh resolutions, as confirmed by the numerical experiment results shown in Figure 10.

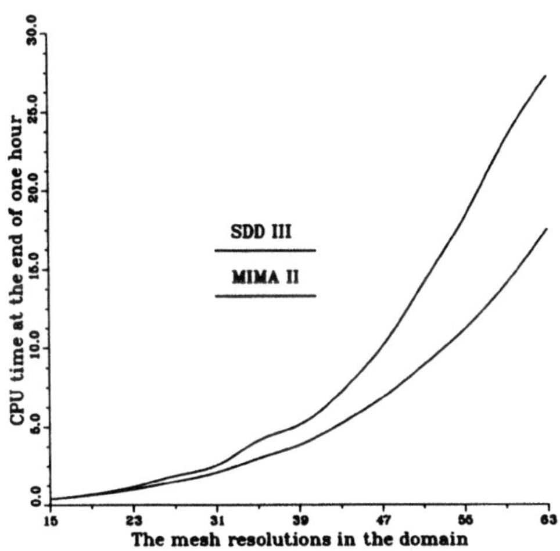

Fig. 10. A comparison in terms of CPU time between the SDD III and MIMA II algorithms for various mesh resolutions, where the horizontal coordinates n, for $n = 15, 19, \ldots, 63$, stand for $n \times n$ mesh resolutions in the original domain.

Finally, we report results of parallel implementation of the MIMA II algorithm. We designed and developed a multitasking code to implement the algorithm MIMA II on the CRAY YMP/432 supercomputer using four vector processors simultaneously in a dedicated environment.

The domain decomposition approach formulated by using the Schur DD method or the MIM approach essentially reduces the solution of a large linear system into that of solving several disjoint smaller subsystems. The resulting parallelism may be exploited at the subroutine level. Calculations involved in the finite element discretization are, however, exploited at the do-loop level. Special attention is paid to calculations of elemental properties and assembly of the local stiffness matrices into a global system matrix (assembly process).

There exists inherent parallelism in the assembly process (each individual element makes its own independent contribution to the global stiffness matrix). The important issue here is, however, to prevent any particular entry in the global matrix from being modified simultaneously. An inter-process synchronization mechanism, the lock, must be invoked in order to guard the corresponding critical region in the code. It should be pointed out, nevertheless, that the critical region involved in the assembly process may be removed by assembling the elemental contributions in a specific order. A multicoloring technique is described and implemented in [5] for removing the critical region and minimizing the number of synchronization points in the assembly process.

The word speed-up is sometimes misleading. Speed-up may be defined to be the ratio of the wall clock time elapsed in a dedicated system, for computing the same problem, using the best serial algorithm and the parallel algorithm. However, the optimal serial algorithm is usually unknown, especially for modeling a meaningful physical process.

Since the best serial treatment of the present problem can not be determined, the speed-up results reported here refer to measurements of wall clock time in a dedicated system relative to the uni- and multi-processor implementation of the same domain decomposition algorithm (see also, among others, [19; 25; 32]). This definition properly incorporates communication overhead and synchronization delays and shows how well the domain-to-processor mappings are done. However, this definition presents a serious drawback. Following this definition, a parallel algorithm achieving a perfect speed-up may actually take longer time to execute than a serial algorithm for solving the same problem.

Having clarified the meaning of the speed-up here, we present graphically the results of the parallel implementation for various mesh resolutions in Figure 11.

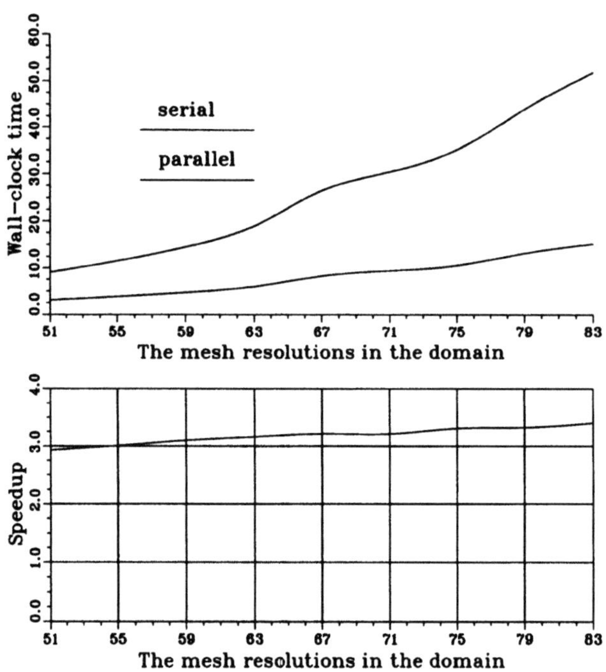

Fig. 11. A comparison of the wall-clock times required for a one hour model integration for the serial and parallel computations by using the MIMA II algorithm and speed-up due to the parallelization of MIMA II vs. the mesh resolutions in the entire domain, where the horizontal coordinates n, for $n = 51, 55, \ldots, 83$, stand for $n \times n$ mesh resolutions in the original domain.

Many recurrence relations which appeared in the original single domain finite element code (see [35]), are not intrinsic and have been removed. For example, the global nodal numbers are fully determined by the local element nodal numbers, once a particular element has been selected. Hence the relations between the global numbering of the nodes, local numbering of the element vertices and element numbering can be calculated independent of the calculations made for the previous elements.

5. Summary and conclusions

In this contribution, we have proposed a MIM non-overlapping DD approach and applied it to the efficient parallel finite element simulation of the shallow water flow. The well-known Schur DD method is also employed for solving the same problem for comparison purposes. Both MIM and Schur DD methods provide logical structures for mapping a whole computational effort to a number of processors for parallel processing.

One of the key ingredients for the Schur DD method is the specification of interface variables. This is done by a divide-and-feedback process. The efficiency of this approach is sensitive to the number of iterations required for convergence on the interfaces and computational costs of the subdomain solvers.

The MIM DD method successively updates the solutions in subdomains and interfaces with increasingly closer initial solutions. The computational costs in both subdomains and interfaces decrease as the iterative procedure continues. Thus it mitigates the disadvantage due to the unavailability of fast subdomain solvers in more complicated cases.

For more realistic and meaningful applications where fast subdomain solvers are not available, the MIM DD method is to be preferred to the Schur DD method.

We plan to explore, in the near future, other possible ways for constructing the modified interface matrix in order to render MIM approach even more efficient and robust.

Acknowledgements

It is a pleasure for the first author to acknowledge the support by US DOE contracts # 120054323 and DE-FC05-85ER250000, through the Supercomputer Computations Research Institute of the Florida State University.

References

[1] H.H. Ahlberg, E.N. Nielson, and J.L. Walsh. *The Theory of Splines and Their Application, Mathematics in Science and Engineering*. 38. Academic Press, New York, 1967.

[2] O. Axelsson and V.A. Barker. *Finite Element Solutions of Boundary Value Problems, theory and computation*. Academic Press, New York, 1984.

[3] P.E. Bjørstad and O.B. Widlund. Iterative methods for the solution of elliptic problems on regions partitioned into substructures. *SIAM J. Numer. Anal.*, 23(6):1097–1120, 1986.

[4] J.H. Bramble, J.E. Pasciak, and A.H. Schatz. The construction of preconditioners for elliptic problems by substructures. *Math. Comp.*, 47:103–134, 1986.

[5] Y. Cai and I.M. Navon. Parallel block preconditioning techniques for the numerical simulation of the shallow water flow using finite element methods. *J. Comput. Phys.*, submitted.

[6] Y. Cai and I.M. Navon. Domain decomposed preconditioners for the finite element shallow water flow modeling. In D.E. Keyes and J. Xu, editors, *Seventh International Conference on Domain Decomposition Methods in Scientific and Engineering Computing*. American Mathematical Society, to appear.

[7] T.F. Chan. Analysis of preconditioners for domain decomposition. *SIAM J. Numer. Anal.*, 24(2):382–390, 1987.

[8] T.F. Chan and D.E. Keyes. Interface preconditioner for domain-decomposed convection-diffusion operators. In T.F. Chan, R. Glowinski, J. Périaux, and O.B. Widlund, editors, *Third International Symposium on Domain Decomposition Methods for Partial Differential Equations*, pages 245–262. SIAM, Philadelphia, 1990.

[9] T.F. Chan and T.P. Mathew. The interface probing technique in domain decomposition. *SIAM J. Matrix Anal. Appl.*, 13(1):212–238, 1992.

[10] T.F. Chan and D. Resasco. A survey of preconditioners for domain decomposition. Technical Report 414, Computer Science Dept., Yale Univ., 1985.

[11] T.F. Chan and D.C. Resasco. A framework for the analysis and construction of domain decomposition preconditioners. In R. Glowinski, G.H. Golub, G.A. Meurant, and J. Périaux, editors, *First International Symposium on Domain Decomposition Methods for Partial Differential Equations*, pages 217–230. SIAM, Philadelphia, 1988.

[12] Q.V. Dinh, R. Glowinski, J. Périaux, and G. Terrasson. On the coupling of viscous and inviscid models for incompressible fluid flows via domain decomposition. In R. Glowinski, G.H. Golub, G.A. Meurant, and J. Périaux, editors, *First International Symposium on Domain Decomposition Methods for Partial Differential Equations*, pages 350–369. SIAM, Philadelphia, 1988.

[13] M. Dryja. A capacitance matrix method for Dirichlet problem on polygon regions. *Numer. Math.*, 39:51–64, 1982.

[14] M. Dryja. A finite element-capacitance method for elliptic problems on regions partitioned into subregions. *Numer. Math.*, 44:153–168, 1984.

[15] R. Glowinski, W. Kinton, and M.F. Wheeler. Acceleration of domain decomposition algorithms for mixed finite elements by multi-level methods. In T.F. Chan, R. Glowinski, J. Périaux, and O.B. Widlund, editors, *Third International Symposium on Domain Decomposition Methods for Partial Differential Equations*, pages 263–289. SIAM, Philadelphia, 1990.

[16] R. Glowinski, J. Périaux, and G. Terrasson. On the coupling of viscous and inviscid models for compressible fluid flows via domain decomposition. In T.F. Chan, R. Glowinski, J. Périaux, and O.B. Widlund, editors, *Third International Symposium on Domain Decomposition Methods for Partial Differential Equations*, pages 64–97. SIAM, Philadelphia, 1990.

[17] G.H. Golub and D. Mayers. The use of preconditioning over irregular regions. In R. Glowinski and J.L. Lions, editors, *Computing Methods in Applied Science and Engineering VI*, pages 3–14. North-Holland, 1984.

[18] A. Grammeltvedt. A survey of finite difference scheme for the primitive equations for a barotropic fluid. *Mon. Wea. Rev.*, 97(5):384–404, 1969.

[19] W.D. Gropp and D.E. Keyes. Complexity of parallel implementation of domain decomposition techniques for elliptic partial differential equations. *SIAM J. Sci. Stat. Comput.*, 9(2):312–326, 1988.

[20] I. Gustafsson. A class of first order factorization methods. *BIT*, 18:142–156, 1978.

[21] I. Gustafsson. Modified incomplete cholesky (mic) methods. In D.J. Evans et al., editors, *Preconditioning Methods: analysis and applications*, pages 256–294. Gordon and Breach, Science Publishers, Inc., 1983.

[22] L.A. Hageman and D.M. Young. *Applied Iterative Methods*. Academic Press, New York, 1981.

[23] D.E. Keyes. Domain decomposition: a bridge between nature and parallel computers. Technical Report 92-44, ICASE, 1992.

[24] D.E. Keyes and W.D. Gropp. A comparison of domain decomposition techniques for elliptic partial differential equations and their parallel implementation. *SIAM J. Sci. Stat. Comput.*, 8(2):s166–s202, 1987.

[25] D.E. Keyes and W.D. Gropp. Domain decomposition for nonsymmetric systems of equations: examples from computational fluid dynamics. In T.F. Chan, R. Glowinski, J. Périaux, and O.B. Widlund, editors, *Second International Symposium on Domain Decomposition Methods for Partial Differential Equations*, pages 321–339. SIAM, Philadelphia, 1989.

[26] D.A. Kopriva. Computation of hyperbolic equations on complicated domains with patched and overset Chebyshev grids. *SIAM J. Sci. Stat. Comput.*, 10(1):120–132, 1989.

[27] D.A. Kopriva. Domain decomposition with both spectral and finite difference methods for

the accurate computation of flows with shocks. *Applied Numer. Math.*, 6:141–151, 1989.
[28] D.A. Kopriva. Spectral solution of inviscid supersonic flows over wedges and axisymmetric cones. *Computers and Fluids*, 21(2):247–266, 1992.
[29] J. Mandel. Two-level domain decomposition preconditioning for the p-version finite element method in three dimensions. *Int. J. Numer. Methods Eng.*, 29:1095–1108, 1990.
[30] J.A. Meijerink and H.A. van der Vorst. An iterative solution method for linear systems of which the coefficient matrix is a symmetric M-matrix. *Math. of Comp.*, 31(137):148–162, 1977.
[31] G. Meurant. Domain decomposition methods for partial differential equations on parallel computers. *Int. J. Supercomputer Appl.*, 2(4):5–12, 1988.
[32] R. Natarajan. Finite element applications on a shared-memory multiprocessor: algorithms and experimental results. *J. Comput. Phys.*, 94:352–381, 1991.
[33] I.M. Navon. Finite-element simulations of the shallow-water equations model on a limited area domain. *Appl. Math. Modelling*, 3:337–348, 1979.
[34] I.M. Navon. A Numerov-Galerkin technique applied to a finite-element shallow-water equations model with enforced conservation of integral invariants and selective lumping. *J. Comput. Phys.*, 52:313–339, 1983.
[35] I.M. Navon. FEUDX: A two-stage, high-accuracy finite-element FORTRAN program for solving shallow-water equations. *Computers and Geosciences*, 13:255–285, 1987.
[36] I.M. Navon. PENT: A Periodic Cyclic Pentadiagonal System Solver. *Communications in Numerical Methods*, 3:63–69, 1987.
[37] I.M. Navon and U. Muller. FESW-a finite-element Fortran IV program for solving the shallow-water equations. *Advances in Engineering Software*, 1:77–86, 1979.
[38] J.M. Ortega and R.G. Voigt. Solution of partial differential equations on vector and parallel computers. *SIAM Rev.*, 27:147–240, 1985.
[39] J.S. Przemienniechi. Matrix structural analysis of substructures. *AIAA J.*, 1:138–147, 1963.
[40] A. Quarteroni, F. Pasquarelli, and A. Valli. Heterogeneous domain decomposition: principles, algorithms, applications. Technical Report UMSI 91/183, University of Minnesota Supercomputer Institute, 1991.
[41] M.M. Rai. A conservative treatment of zonal boundaries for Euler equation calculations. *J. Comput. Phys.*, 62:472–503, 1986.
[42] Y. Saad and M.H. Schultz. GMRES: A generalized minimal residual algorithm for solving nonsymmetric linear systems. *SIAM J. Sci. Stat. Comput.*, 7(3):856–869, 1986.
[43] H.A. Schwarz. Über einige Abbildungsaufgaben. *Ges. Math. Abh.*, 11:65–83, 1869.
[44] P. Sonneveld. CGS, A fast Lanczos-solver for nonsymmetric linear systems. *SIAM J. Sci. Stat. Comput.*, 10(1):36–52, 1989.
[45] R. Temam. Survey of the status of finite element methods for partial differential equations. In D.L. Dwoyer, M.Y. Hussaini, and R.G. Voigt, editors, *Finite Elements, theory and application*, pages 1–33. Springer-Verlag, New York, 1988.
[46] C. Temperton. Algorithms for the solution of cyclic tridiagonal systems. *J. Comput. Phys.*, 19:317–323, 1975.
[47] H.A. van der Vorst. Bi-CGSTAB: a more smoothly converging variant of CG-S for the solution of nonsymmetric linear systems. *SIAM J. Sci. Stat. Comput.*, 13(3):631–644, 1992.
[48] R.S. Varga. *Matrix Iterative Analysis*. Prentice-Hall, Englewood Cliffs, N.J., 1962.

PARALLEL COMPUTING AND OPERATIONAL SEA-ICE FORECASTING

Jeremy Cook
Parallab, Dept. of Informatics
University of Bergen
N-5020 Bergen, Norway
Jeremy.Cook@ii.uib.no

ABSTRACT. In this paper we review current state-of-the-art hardware facilities for application to sea-ice modelling. We also discuss data sources and future trends to emphasise the current, and indeed, future requirements for massively parallel supercomputing resources.

1 Introduction

Operational forecasting is a computationally intensive task performed on potentially large data sets. NERSC (Nansen Remote Sensing Center) is responsible for operational forecast of sea-ice formation in the Barents Sea for example, there are many data sources to be considered, producing data at a high rate. Data is obtained from diverse sources and is obtained at different times. The operations involved include processing of raw Synthetic Aperture Radar (SAR) data, post processing to determine the ice kinematics and ice classification, and modelling to determine ice movement. In this paper, we consider the various aspects of operational forecasting, and the impact of parallel supercomputers in this field.

We will begin by giving a brief review of a number of systems which we have available for scientific computation. Parallab, the laboratory for parallel computations, is the result of parallel computing activities in Bergen going back as far as 1985. Parallab is based at the Department of Informatics at the University of Bergen in Norway. Access to state of the art hardware facilities is provided to our customers without the need for substantial hardware investments. The facilities currently available (October 1993) are shown in table 1.

These systems are integrated into the department's FDDI network, which reaches to all desktops, and additionally, into the national 34Mb/s SUPERNETT network established by the Norwegian Supercomputing Committee. Within the University, a highspeed central GIGAswitch from DIGITAL equipment ensures high bandwidth availability between all systems.

F.-X. Le Dimet (ed.), High Performance Computing in the Geosciences, 107–113.

Name	Number of nodes	Realistic peak	Total memory
Intel Paragon	100	4.0 Gflop	3.0 Gbyte
MasPar MP-2	16384	2.0 Gflop	1.0 Gbyte
Sparc-10 cluster	50	1.5 Gflop	1.6 Gbyte
DEC α-cluster	15	1.5 Gflop	1.0 Gbyte
IBM-RS6000 cluster	7	0.5 Gflop	1.0 Gbyte
Total	16556	9.5 Gflop	7.6 Gbyte

Table 1: The range of parallel and distributed systems available at parallab

A well balanced parallel computer should have at least one byte of memory per double precision flop. A well balanced parallel computer must also have a communication bandwidth that increases with the number of processors. The systems in the above table are fairly well balanced, offering large memories, and high bandwidth communications. The MasPar MP-2 could be considered a little less well balanced than its predecessor the MP-1 as the router and nearest-neighbour communications remain the same for the new generation system. However, the MP-2 system does offer increasing communication bandwidth as the number of processors increases and the access time to nearest neighbours is of the same order as the time for one double precision operation.

We can also note that within about 12 months from the date of writing, *all* these systems should be able to run the same program written in *HPF* (High Performance Fortran [Hig93]). Apart from the MasPar system, all of these systems can run the same program using explicit message passing based on PVM (Parallel Virtual Machine [GBD+93]).

From table 1, we see that these systems have widely varying capabilities. The Intel Paragon system is equipped with some 100 commodity processors. The processors in this case are *tightly coupled* with a proprietary router network. The MP-2 system is equipped with a large number of slower proprietary processors using a high-speed proprietary router network. The advantage of using proprietary CPUs in this case outweigh the advantages of commodity RISC units. 32 processors are squeezed onto each chip at low cost but the large number of such processors gives a high-performance system.. The remaining cluster systems are all based on generally available workstation technology and high-speed networking technology (FDDI). These systems are attractive for many sites in that they allow otherwise idle workstations to take part in compute intensive operations, however such systems are much more loosely coupled and offer somewhat lower performance gains for general applications than the tightly coupled systems.

In our table we give figures for "Realistic peak". By this we mean the peak performance that can be realistically obtained in an application.

2 SAR, Sea Ice and Data Assimilation

There are a number of factors that lead us to the conclusion that parallel computers are the *only* cost effective solution to the future processing needs for processing of SAR data, sea-ice modelling and other areas related to operational forecasting. In our area of interest, operational forecasting, we acquire data from many different sources. Let us first review the nature of the problem of sea-ice monitoring.

There is an interest in sea-ice variability in three timescales. Typically 0—7 day intervals for operational forecasting, 3—6 months for prediction of freeze-up/breakup, and interannual observations for global climate change prediction.

The monitoring of sea-ice conditions is required for ice charts (weekly), forecast initialisation, and global change detection. The most commonly available data sources for sea-ice information are satellite bases and buoys.

2.1 Satellite Data Sources

There are a number of satellite sources available to provide information about the nature of the sea surface and sea-ice.

2.1.1 Passive microwave (DMSP SSM/1) The passive microwave satellite offers a 25 km resolution, enabling detection of the ice edge to within approximately 10 km. Daily global all weather coverage is offered. Data is available in "real time", *ie.*, 6 hours after the pass is completed.

Passive microwave data can be used for 1st year and multi-year sea-ice classification, and to determin ethe concentration of sea-ice.

2.1.2 Infra-red, visible bands (NOAA AVHRR) This class of satellites offers resolution down to 1km. Daily global coverage can be obtained, howeverinfra-red and visible band radiation cannot penetrate cloud. Visible band cannot be used at night.

Infra-red and visible data are used to determine the ice edge, extent and velocities.

2.1.3 Synthetic Aperture Radar (SAR) Current SAR satellites provide resolution down to 25m, where each scene is 100 × 100km. Orbit repeat occurs every 3-35 days with overlap. All weather and night time coverage are available.

SAR satellites are used operationally (used actively in operational forecasting at NERSC). Uses include determining ice velocity (kinematics), ice classification (histogram thresholding) and feature identification (ice edge, leads, ridging, wave influences).

In addition to satellite sources, data is also available from other sources such as buoy data and data from research vessels.

Data from a single source such as SAR data cannot give reliable classification of sea-ice and water. Coverage is incomplete, and is not from the same point in time for all scenes. In order to derive a complete picture, data assimilation is required to assimilate data from many sources taken from various times and places.

2.2 Unending Appetite for Processing Power

The current SAR satellites – ERS-1 and the Japanese JERS-1 – provide a swath of data 100km wide. For a 100 × 100km scene, as much as 250 MBytes of raw data are produced. By 1995 next-generation satellites will be on-line, incluting the US/Canadian Radarsat, the European ERS-2 and the Japanese JERS-2. These satellites will be capable of producing scenes of 500 × 500km with the same or better resolution as the current day satellites, giving a subsequent increase in the amount of raw data to be received and archived.

3 SAR Processing

SAR image formation is a computationally intensive task performed on large data sets. There is some interest in building special purpose computers to process SAR images. In the author's view, general purpose parallel computers are better suited to this task, and provide better value for money. Mastin *et al.* [MPG93] review the suitability of the Connection Macine CM-2, The nCube-2 and the Intel iPSC/860 for this task with a favorable conclusion. In our work we have used the MasPar MP-1 and MP-2 computers to process raw SAR data from the ERS-1 satellite. We find that the architecture of the MP-2 to be highly suited to SAR processing, and real-time[1] image formation is possible.

3.1 Basic SAR Concepts

There are a number of schemes for implementing the processing of SAR images from RAW data. In the algorithm described here, the processing is divided into three parts: range compression, azimuth compression and auxiliary processing. The range and azimuth compression stages constitute the main image formation process. Auxiliary processing consists of doppler centroid estimation, generation of compression filters and geometric correction of the final image.

Before processing can begin, doppler centroid estimation and compression filter generation must be carried out. However the data generated is static for a given satellite configuration and filter generation is carried out off-line.

3.1.1 Range Compression During range compression the raw data vector (5bit complex data) is read in, unpacked and converted to complex single precision vectors. The complex vectors are then Fourier transformed, multiplied with the compression filter (already in the spectral domain). After demodulation, the vector is inverse Fourier transformed to give the compressed pulse echo in the time domain.

3.1.2 Azimuth Compression For azimuth compression the range compressed data is accessed in the azimuth direction. Again it is Fourier transformed. It is then filtered and inverse Fourier transformed reducing the bandwidth by a factor 4.

[1] By real-time, we mean that we can process the data much faster than the transmission takes from the satellite to ground station.

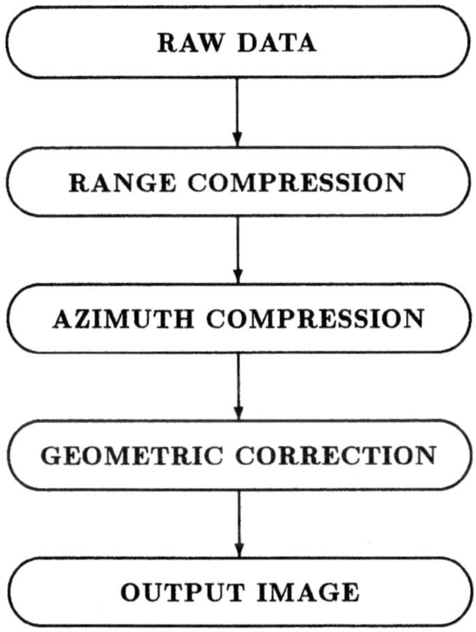

Figure 1: Overview of the basic steps in SAR processing

Before the final image is produced, some geometric correction and image resampling is performed.

We find that all of the stages described here are well suited to parallelisation. All of the stages can be computed in parallel. The major hurdle to overcome is the large amount of core memory required to hold all of the data, and the large I/O throughput which is required if I/O is not to be a bottleneck in the process.

Our first implementation used a MasPar MP-1 computer with 8192 processors. The first generation MasPar system was equipped with 16Kbytes of RAM per processor. Although this amounts to 128Mbytes for the whole machine, it was clear from the start that this would not be enough memory to hold a 1024 element range vector of complex single precision data and several temporary vectors of the same size. In our first implementation we therefore shared each range vector between two PEs. Even numbered PEs held even numbered range vector elements and odd numbered PEs held odd numbered range vector elements. All work was carried out by the pairs of PEs with an exchange of data occuring whenever a shift of an odd number of elements was required. The programming effort was complicated by this fact, and a number of extra nearest neighbour communications were incurred. Despite the extra programming complexity caused by restricted memory, the work was completed in just 3 weeks and we obtained an execution time well inside our goal (see table 2). Table 2 shows timings for the dominant parts of the code.

[2]This is the total time as measured, extra overhead not measured in the individual timings is taken into

Code section	8192 PE MP-1	16384 PE MP-2
Reading raw data	72.14	27.83
Range Compression	126.43	36.49
Azimuth Compression	52.73	18.93
Geometric Correction	103.40	22.94
Writing image	71.69	29.73
Total execution time[2]	438.00	146.49

Table 2: Performance (times in seconds) for full 26000 line SAR image. The MP-1 and MP-2 figures both use the "small memory model" discussed in the text.

In these figures, I/O is 30—40% of the time. If I/O could be carried out asynchronously, and overlapped with communications, we estimate that all I/O time in table 2 would be virtually eliminated. This would save about 2.5 minutes for the MP-1 case and about a minute in the MP-2 case giving us processing times of 5 minutes and 1.5 minutes respectively. The current release of the operating system has features to allow asynchronous I/O operations, however a large amount of I/O ram buffer memory is required, and this is not currently available on the system at **para//ab** to allow us to carry out testing.

The MasPar MP-2 system at **para//ab** has a superior I/O throughput to the MP-1 with up to 18Mbyte/sec attainable. The MP-2 system (and later versions of the MP-1) is nolonger restricted to 16Kbyte per processor. 64Kbyte is now available as standard (*ie.*, 1Gbyte core memory for the full 16K PE system). For SAR processing, this means that we are no longer required to share range processing between two PEs.

In table 3 we show the execution times for a full SAR image using a "large memory model". The algorithm in this case does not need to share memory between two processors as we were required to do in the MP-1. This allows us to carry out FFTs more efficiently as no communication is required for FFTs in the azimuth direction. In addition, the large memory model allows for more efficient I/O – the data can be accessed the data in the raw data file sequentially rather than random access as was required in the MP-1 case.

Code section	8192 PE MP-2
Reading raw data	43.58
Range Compression	49.48
Azimuth Compression	27.90
Geometric Correction	43.66
Writing image	30.92
Total execution time	204.37

Table 3: Performance (times in seconds) for full 26000 line SAR image using a *large memory model.*

account.

References

[GBD+93] Al Geist, Adam Beguelin, Jack Dongarra, Weicheng Jiang, Robert Manchek, and Vaidy Sunderam. Pvm 3 user's guide and reference manual. Technical Report ORNL/TM-12187, Oak Ridge National Laboratory, May 1993.

[Hig93] High Performance Fortran Forum. High performance fortran language specification. Technical report, Rice University, January 1993.

[MPG93] Gary A. Mastin, Steven J. Plimpton, and Dennis C. Ghiglia. A massively parallel digital processor for spotlight synthetic aperture radar. *International Journal of Supercomputer Applications*, 7(2):97–112, 1993.

COUPLED OCEAN-ATMOSPHERE SIMULATIONS

L.TERRAY, and O.THUAL
Climate Modelling and Global Change Project
Cerfacs
42 Avenue Coriolis
31057 Toulouse Cedex
France

ABSTRACT.

Climate modelling for the prediction of anthropogenic global change requires to take into account the coupling phenomena between the ocean and the atmosphere. Several approaches, ranging from simple dynamical systems studies to full size coupled General Circulation Models (GCM) are necessary to apprehend the basic physical processes underlying the climate mean state and its seasonal and interannual variability. In this paper, we present the environment and the strategy we have used to realize some coupled ocean-atmosphere simulations with a global atmospheric GCM and a regional oceanic one restricted to the Pacific Ocean. First, we describe the coupling fields by which the interaction mechanisms at the air-sea interface arise. Using a simple 1-D heat equation, the choice for a particular set of boundary conditions is shown to be related to vertical discretization. Then we detail the design of the coupling software we have written in order to do the simulations. We present our general strategy which is based on coupling fluxes analysis, forced and coupled simulations and we show some results from our Pacific coupling experiments. Last, we conclude by pointing out some open research axes focusing on global coupling including sea-ice and vegetation.

1. Introduction

During the past ten years, the understanding of climate has considerably progressed. One important explanation lies in the development of observations, through specific campaigns and monitoring programs, principally with satellite data. But the exploitation of theses observations is greatly enhanced by climate modelling, ranging from simple Energy Balance Models (EBM) to Coupled General Circulation Models (CGCM). This last approach, which is the reference for simpler modelling, has taken advantage of the tremendous increase of supercomputing power but also of a better understanding of the physics and dynamics involved within the atmosphere-ocean system.

Predicting anthropogenic climate change must been preceded by an accurate modelling of the present climate. It is well settled now that this activity must take into account the coupling between the atmosphere, the ocean and the sea-ice. Two important phenomena depend strongly of this coupling and are intensively studied

F.-X. Le Dimet (ed.), High Performance Computing in the Geosciences, 115–123.

by the international community. The first one is the El Niño and Southern Oscillation (ENSO) phenomenon, which involves ocean-atmosphere tropical dynamics, and explains most of the 2 to 10 year interannual variability in these regions. On longer time scales, the climate variability is characterized by interdecadal oscillations which involve intricate feedbacks between the thermohaline circulation, responsible for deep water formation at high latitudes, the sea-ice and the hydrologic cycle in the atmosphere.

The study of these two paradigms involving ocean-atmosphere interactions represent the main challenges for the modelling community before undertaking global change climate predictions on time scale of the order of a century. In order to adress these questions thoroughly, CGCM are required as they include the complete set of physical processes essential for the comprehension of ocean-atmosphere energetics. Furthermore they provide a consistent basis to analyze the physics and thermodynamics at work within each component.

2. Air-Sea interface

Most of the climate models which couple an atmospheric general circulation model (AGCM) and an oceanic circulation model (OGCM) adopt the following scheme: at each coupling time step, several fluxes (heat fluxes, wind stresses and evaporation minus precipation), calculated by the AGCM are passed to the OGCM, and the sea surface temperature (SST) and possibly surface currents, calculated by the OGCM, are passed to the AGCM. If a sea-ice model is embedded in the OGCM, the sea-ice extent (SIE) needs to be passed as well. Otherwise it is prescribed from climatology.

The basis for such procedure can be explained simply using the following framework. A simple model which can be used to study why fluxes are passed from the atmosphere to the ocean, and SST from the ocean to the atmosphere is the 1D heat equation:

$$c(z)\partial_t T = \partial_z \left[\lambda(z)\partial_z T\right] , \tag{2.1}$$

where $T(z,t)$ is the temperature, $c(z)$ a vertical profile of heat capacity and $\lambda(z)$ a vertical profile of diffusivity. For the boundary conditions, we assume that the heat flux is zero, $F(z) = \lambda\partial_z T = 0$ at the top $z = a_2$ and the bottom $z = -a_1$, where a_1 and a_2 are positive lengths defining the vertical interval $[-a_1, a_2]$. We suppose that we divide this interval into two subdomains $[-a_1, 0]$, called the oceanic domain, and $[0, a_2]$, called the atmospheric domain.

Even if there is a discontinuity of $c(z)$ or $\lambda(z)$, e.g. at the ocean-atmosphere interface $z = 0$, the temperature $T(z)$ and the heat flux $F(z) = \lambda\partial_z T$ must be continuous in order to be solution of the heat equation (2.2.1). From this observation, we can show mathematically that the boundary value problem of the heat equation (2.2.1) on the domain $[-a_1, a_2]$, with the no-flux boundary conditions at the top and bottom, is equivalent to the union of two boundary value problems. These are the heat equation:

$$c_1(z)\partial_t T_1 = \partial_z \left[\lambda_1(z)\partial_z T_1\right] , \tag{2.2}$$

restricted to the domain $[-a_1, 0]$ with the no-flux at the bottom $z = -a_1$ and a boundary condition B_1 at the ocean-atmosphere interface $z = 0$, coupled to the heat equation:

$$c_2(z)\partial_t T_2 = \partial_z \left[\lambda_2(z)\partial_z T_2\right] , \tag{2.3}$$

on the domain $[0, a_2]$ with a boundary condition B_2 at $z = 0$ and the no-flux boundary condition at the top. In these equation, c_1, c_2, λ_1 and λ_2 are the restrictions of $c(z)$ and $\lambda(z)$ on their respective domains.

The coupling boundary conditions B_1 and B_2 can be chosen independantly to be either the Dirichlet boundary condition D:

$$T_2(0) - T_1(0) = 0 \tag{2.4}$$

or the Neuman boundary condition N:

$$F_2(0) - F_1(0) = \lambda_2(0)\partial_z T_2(0) - \lambda_1(0)\partial_z T_1(0) \tag{2.5}$$

or any linear combination of them. For instance, we can choose the Neuman boundary condition for the ocean, i.e. $B_1 = N$, and the Dirichlet boundary condition for the atmosphere, i.e. $B_2 = D$. But the reverse can also be chosen, or the same boundary condition for the two media. The proof of this statement is due to the unicity of the solution of the boundary value problem on the whole domain, given an initial condition.

The reason why a Neuman boundary condition is used for the oceanic model, and a Dirichlet boundary condition is used for the atmospheric model in most of the coupled GCM, is simply due to the vertical discretization. This can be simply seen by looking at the finite difference scheme of the heat equation (2.2.1):

$$c_j \frac{T_j^{n+1} - T_j^n}{\delta t} = \frac{F_{j+\frac{1}{2}}^n - F_{j-\frac{1}{2}}^n}{\delta z} \tag{2.6}$$

with $F_{j+\frac{1}{2}}^n = \frac{T_{j+1}^n - T_j^n}{\delta z}$ and $F_{j-\frac{1}{2}}^n = \frac{T_j^n - T_{j-1}^n}{\delta z}$ where $T_j^n = T(j\delta z, n\delta t)$ is the discretize approximation of the temperature field. We see that the discretized fluxes $F_{j+\frac{1}{2}}^n$ are defined on the half integer grid points $(j + \frac{1}{2})\delta z$.

When splitting the whole domain $[-a_1, a_2]$ into two domains $[-a_1, 0]$, the ocean, and $[0, a_2]$, the atmosphere, several choices are possible for the discretisation of the two coupled boundary value problems (2.2.3) and (2.2.4) with the boundary conditions B_1 and B_2. If the ocean-atmosphere interface $z = 0$ is chosen as an half integer grid point, it would be appropriate to chose B_1 and B_2 to be the Neuman boundary condition (2.2.5), i.e. to impose the continuity of the fluxes. If this interface is an oceanic integer grid point, it would be appropriate to chose B_2 to be the Dirichlet boundary coundition (2.2.6) and B_1 to be a Neuman boundary imposed at the first half integer grid point above the interface $z = 0$. This last situation is the one encountered for most of the OGCM/ACGM couplings.

3. OASIS: a coupler for climate simulations

OASIS [1] is a set of fortran routines which enables the coupling of oceanic and atmospheric models, for instance general circulation models (GCM) or simpler ones. In the following, we use GCM as a generic term for an ocean or an atmospheric model whether it is a real GCM or a simpler model. OASIS main concepts are the following:

* The three programs are considered as different processes in the unix sense and so synchronization is done using the message passing concept relying either on standard unix tools such as pipes or signals or on the parallel virtual machine (PVM) software [2]. The GCM and the coupler run concurrently interactively or in batch mode on a supercomputer.

* The coupler does all the coupling which is basically to provide boundary conditions for each model on its own grid when they are needed. The only job the models have to do is to put under a given format the data which need to be transferred to the coupler. Conversely they must be able to read the data they need (their own boundary conditions) coming from the coupler. Fortran units connected with files shared by the coupler and the two models are used to exchange (i.e read and write) data between the three programs. The common format can be either simple binary or a format specific to the GCM. When PVM is used, data is transferred within PVM and it is then possible to have the models running on different machines.

* The package is extremely modular, each routine has a well defined task separated from the rest of the code. For instance, there are in the actual version several interpolation packages (Lagrange, splines, nearest-neighbor ...) and one can switch between these different methods by changing namelist parameters.

Figure 1 shows a schematic picture which illustrates the concurrent evolution of the 3 programs.
This framework has important implications:

* Any atmospheric model can be coupled to any oceanic model via OASIS with very few changes in both models. In particular, it is possible to use a dataset as a passive atmospheric model to force different OGCM for intercomparison purposes.

* Other interpolation packages can be used as long as they have a similar calling sequence.

* Models structure is entirely preserved (i.e multitasking options, I/O scheme...). There is no constraint on the computing aspect (i.e for instance on the number of CPU devoted to each program)

* There is no "hidden variables"; everything is written in fortran following the DOCTOR norm from ECMWF [3] and can be understood quickly by the user.

A complete description of OASIS can be found in [1]. OASIS has been used extensively in forced or coupled simulations which have shown its robustness and flexibility (see Section 4).

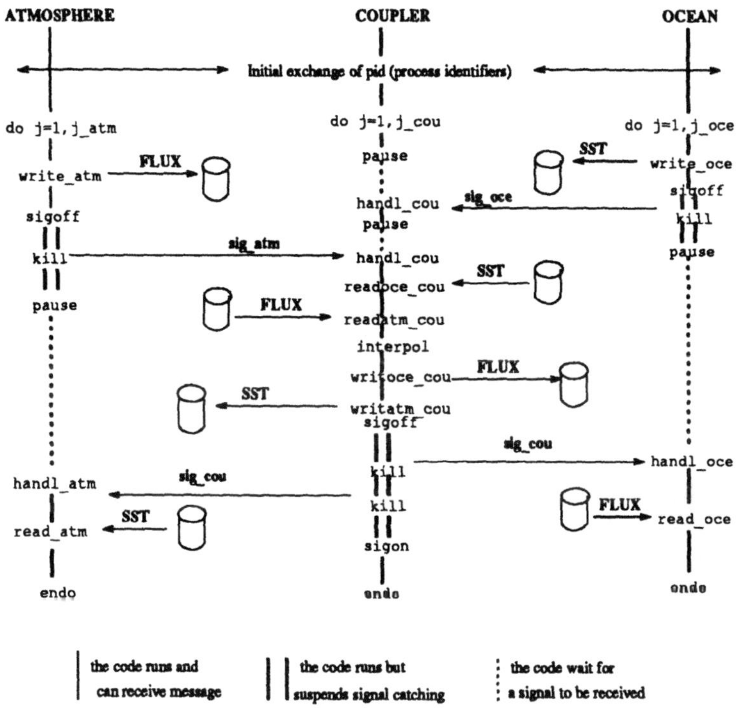

Fig. 1. OASIS coupling scheme

4. Strategy for climate studies

The two GCM used in this work are the components of the so-called French Community Climate Model which aims to be a comprehensive and global system for climate change studies. The atmospheric component is the climate version of the Météo-France global forecasting model ARPEGE [4]. The oceanic one is the Laboratoire d'Océanographie DYnamique et de Climatologie (LODYC) oceanic model OPA which, in this work, is dynamically active over the tropical Pacific [5].

In a first stage, the ARPEGE model has been forced with 10 years of observed SST (Sea Surface Temperature) as boundary conditions for the Atmospheric Model Intercomparison Project (AMIP) period (1979-1988). Then, the five surface fluxes (zonal and meridian components of the wind stress, fresh water flux, total heat flux and solar heat flux) have been used to force the Tropical Pacific version of the oceanic model OPA during this 10 year period.

The numerous data provided by this decadal simulation have been analyzed to study the interannual tropical variability of the Pacific ocean, a low-frequency variability caused by complex interactions between the atmosphere and the ocean which is expressed in an irregular interannual fluctuation between warm El Niño and cold La Niña states. This preliminary run provided very promising results:

the simulated fields are in very good agreement with the available data, and the behavior of the tropical Pacific ocean is correctly reproduced [6]. The comparison between the simulated SST averaged on the full period 1979/1988 and the climatology shows relatively small differences in the Pacific, except for the eastern part of the equatorial Pacific and along the peruvian coasts where the simulated SST is too warm (Figure 2).

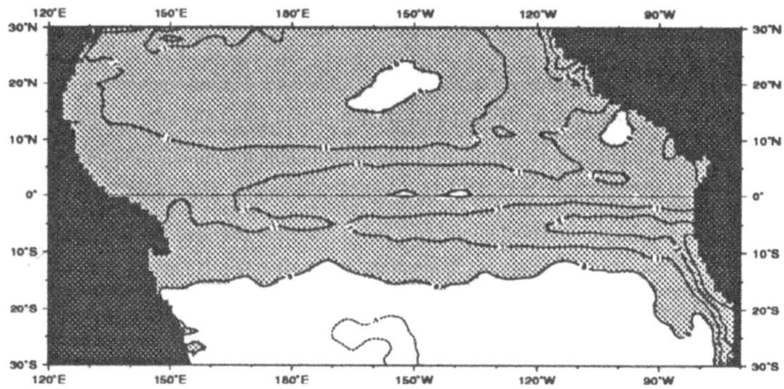

Fig. 2. Mean SST difference with climatology

The difference in the eastern part of the basin is essentially due the contribution of the two major warm events of 1982-83 and 1986-87. The simulated seasonal cycle is also in good agreement with the observations. At last, the oceanic model succedeed in reproducing the main features of the 1983 El Ninõ event which was the warmest event in the century. Figure 3 represents the SST anomaly over the Pacific basin for February 1983. The central and eastern Pacific show strong positive anomalies and the maximum (4°C) is located around 130°W.

This ten year simulation — a necessary step before the coupling with ARPEGE, and a first validation of the coupler OASIS into an operational mode —, constitutes now a reference run, which will be very much needed to understand and analyze coupled model results.

We have performed recently the first coupling experiments [7] between ARPEGE and OPA with the OASIS coupled on a 10 year period, in order to simulate the mean state, the seasonal cycle and to study potential interannual variability in our coupled system (Figure 4). The two GCMs interact only over the Pacific ocean in the region 40S-48N. The ocean model is forced by the surface wind stress, the heat fluxes (total and solar), and the fresh water flux simulated by the atmospheric

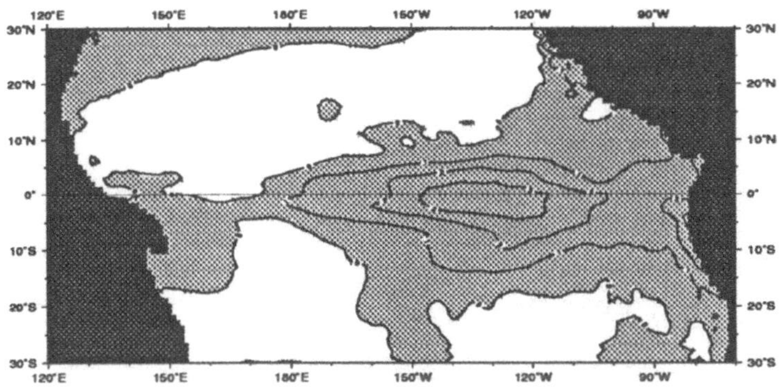

Fig. 3. SST anomaly for February 1983

model, which in turn is forced by the sea surface temperature (SST) simulated by the ocean model. Outside the coupling region, boundary conditions are prescribed from climatology. The deep ocean is relaxed towards the Levitus climatology. Information is exchanged between the two models every day (i.e, every 24 and 96 time steps respectively for the OGCM and the AGCM). The only external forcing is the seasonally varying solar radiation. The oceanic initial conditions are those simulated by a 10-year uncoupled simulation of the Pacific ocean forced with ARPEGE fluxes coming from an AMIP simulation of the ARPEGE model forced with observed SST's. The atmospheric conditions were taken from this same AMIP run. The initial date is July first 1985 and the simulation continued for ten years. The results of this 10 year coupled simulation are very encouraging, and it is part of an intercomparison between several coupled models worldwide, including the effort the Max Planck Institute and the Hadley Center. The ARPEGE/OPA simulation exhibits good results as far as the mean state and the seasonal cycle are concerned. It also shows definite evidence of interannual variability.

Figure 4 shows a longitude-time diagram of the dynamic height anomalies over the whole simulation period. This field represents the heat content in the ocean upper layers. There is a clear oscillation with a 3-year period which is related to atmospheric variability. However the signal amplitude is rather weak, but no coupled CGM has succeeded, so far, to combine realistically the mean state, seasonal cycle and interannual variability.

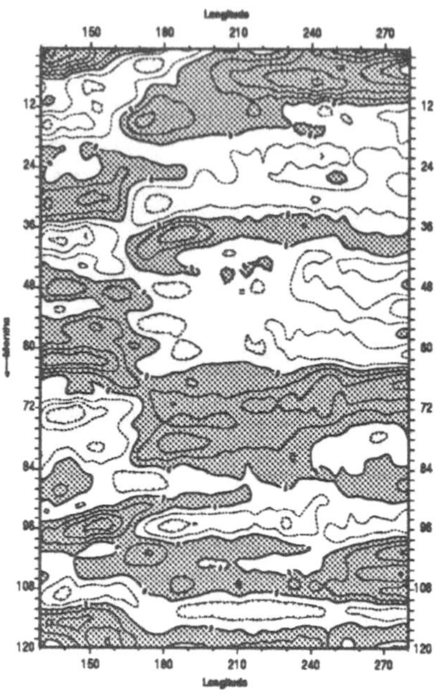

Fig. 4. Time-longitude diagram of dynamic height anomalies over the 10 years

5. Conclusion

In order to couple any two General Circulation Models (Atmosphere and Ocean) we have developed a software based on message passing. We have used this coupler, called OASIS , to perform coupled experiments with ARPEGE and OPA. For the first time in the French climate community, we have successfully coupled ARPEGE and the Tropical Pacific version of OPA on a 10 year period. This coupled simulation gives a correct description of the seasonal cycle despite a warm bias in the ocean mean state. Interannual variability is present although its amplitude is rather weak.

In parallel, we build all the tools necessary to launch experiments with the global version of the oceanic model. In particular, a sea-ice model will be implemented between the ocean and atmosphere in order to describe accuratly deep water formation at high latitudes. We plan not to use any kind of flux correction techniques in order to study carefully the drifts that might occur in a multidecadal simulation. Eventually, these efforts will lead us to climate experiments in which the greenhouse gases are progressively increased according to the currently accepted scenarios. These simulations will be then compared to similar ones performed in a growing number of international climate communities.

References

[1] Terray L (1993) The OASIS Coupler User Guide Version 1.0, *CERFACS Technical Report*, CERFACS **TR/CMGC/93-33**.

[2] Beguelin A, Dongarra J, Geist A, Manchek R, Sunderam V (1991) A users' guide to PVM Parallel Virtual Machine, *Technical Report* ORNL/**TM-11826** Oak Ridge National Laboratory.

[3] Gibson JK (1982) The Doctor system - A DOCumentary ORiented programming system, *ECMWF Technical Memorandum* No. **52**.

[4] Déqué M, Dreveton C, Braun A, Cariolle D (1994) The climate version of Arpege/IFS: a contribution to the French community climate modelling. *Climate Dynamics* (in press)

[5] Delecluse P, Madec G, Imbard M, Levy C (1993) OPA version 7 Ocean General Circulation Model reference manual. *LODYC, internal report* **93/05**

[6] Terray L, Belamari S (1993) Simulation of the 1986-1987 El Niño and 1988 La Niña events with a high resolution tropical Pacific ocean general circulation model. *CERFACS Technical Report*, CERFACS **STR/CMGC/93-30** (to be submitted in *Journal of Physical Oceanography*)

[7] Terray L, Belamari S, Thual O, Deque M, Dandin P, Delecluse P, Levy C (1994) Climatology and interannual variability simulated by the ARPEGE-OPA coupled model. Submitted to *Climate Dynamics*.

First Considerations about Modelling the Ocean General Circulation on MIMD Machines by Domain Decomposition Method

Marc Guyon, and Michel Chartier
CETIIS
24 Boulevard Paul Vaillant-Couturier
F-94200 Ivry-Sur- Seine
France.
Francois-Xavier Roux
ONERA/DI/Division Calcul Parallele
BP 72
29 Av de la Division Leclerc
F-92320 Chatillon
France.
and
Philippe Fraunie
LSEET
Universite de Toulon
BP 132
F-83957 La Garde Cedex
France.

ABSTRACT.

In this paper, we talk about an action of parallelization in Geosciences field.
As a candidate we have chosen the OPA7 code, a three-dimensional primitive-equation finite-difference ocean general circulation model.
We present our point of view about Domain Decomposition Methods, architectures of machines and their programming models.
After we present our choices to parallelize the explicit part (data management, subdomain topology...) and the implicit part (a Domain Decomposition solver).

Introduction

In recent years, we have observed a renewed interest in Domain Decomposition Methods. They allow codes efficiently using the parallel architectures of current scientific computers to be defined. In Earth Sciences field, scientists need to obtain better understanding of physical processes. They need for more complex modelizations. They would like to use much more CPU time for climate experiments for instance and they want a better resolution of mesoscale phenomena in Oceanogra-

F.-X. Le Dimet (ed.), High Performance Computing in the Geosciences, 125–140.

phy field.
Now, technological limits of traditional vector machines seem to have been reached for calculation and memory capacity.
Parallelism is likely to be the new technology to obtain Teraflops and Teraoctets for such a modelization (Adam and Alessandrini, 1994).
Obviously, Earth Sciences is one of the best fields to exploit the potentials of parallel computing. We can note there is a large number of parallelization actions in this context, some of these are applied to a Shallow Water Equations (McBryan, 1992; Navon and Cai, 1992), others concern a simplified model (Clare and Steven, 1993) or more complex model in SIMD field (Wolters, 1992).
In the same way, we have to take advantage of today's (and future) development of scientific supercomputers. In this paper, we present different considerations, in first analysis field, about parallelism in Earth Sciences field. These considerations lead us to adopt Domain Decomposition approach and message passing programming model to exploit MIMD machines potential.
As a candidate for domain decomposition method, there is, in the first part, a desciption of OPA7. We present after our interest for Domain Decomposition Approach and considerations about architectures and programming models. In the third part, we talk about dependency analysis, interface management implementation like Dirichlet conditions and an efficient topological splitting (pencil) applied to the explicit part of OPA7 and a new strategy to solve the implicit one. To conclude, we present some hopefull results on scalability tests and comparisons with CRAY machines concerning the Baroclinic part of OPA7.

OPA7 model

As a candidate to geophysical application of Domain Decomposition methods, the OPA7 three-dimensional primitive-equation finite-difference ocean general circulation model, developed in France by the Laboratoire d'Ocanographie DYnamique et de Climatologie (Chartier, 1986; Madec et al, 1990; Madec et al, 1991; Marti et al, 1992), has been adopted.
The code solves the oceanic three-dimensional Primitive-Equations on the earth surface. The basic assumptions of the model are the Boussinesq and hydrostatic approximations.
A mixing algorithm restores instantaneously the static instability over the whole water column when needed. The rigid-lid technique is used and the baroclinic and barotropic part of the circulation are computed separately. The baroclinic circulation is calculated through an explicit algorithm whose parallelization is presented here. The barotropic circulation (deriving from a stream function in theOPA-7 code) is an implicit phase solved, in the scalar code, by a preconditioned conjugate gradient method.
Finite-difference methods associated to a tensorial formalism are used to solve the equations on a so-called "Arakawa-C" (fully staggered) grid (Arakawa and Messinger, 1976). The reason for staggering the grid is to remove computational modes in the finite difference formulation. OPA7 is a very efficient code, that has already been parallelized in a CRAY multitaskingenvironment (Andrich, Madec

and L'Hostis, 1988). Analysis of OPA7 code had provided us a concrete support to answer the question of parallelizing of such a complex model. In particular, we have to consider in a different way the Baroclinic Flow (explicit phase) and the Barotropic Transport (implicit one).

Domain Decomposition Methods : principle and advantages

Domain decomposition methods have been developed for solving partial differential equations. These methods are especially suited for parallel computing and particularly for distributed memory MIMD machine architecture. This study aims at exploiting domain decomposition methods in geoscience field in order to take advantage of present (and future) development of scientific supercomputers.
The basis of these methods consists in splitting the large computation domain of a numerical experiment into several smaller domains and in solving the set of equations by addressing "independent" local problems.

These methods present considerable advantages :
1. They save absolute time due to parallel processing, as well as CPU time by performing computations on small domains (by reduction of bandwith in matrix factorization case...).
2. They account for the non-uniformity of the physical processes (different parameter values, or algorithms, or parametrization, or even different numerical models can be implemented in each subdomain).
3. They provide a natural fieldwork for zooming methods; this question is complex due to conservative laws for the Baroclinic Flow and continuity conditions for the Barotropic Transport.

So, we think to take advantage of parallel computers and physical potential of Domain Decomposition Approach. We have chosen to develop a Domain Decomposition version of an Ocean General Circulation Model.

Which architecture?

We can divide parallel machines into two main categories :
1. Machines with a small number of processors (Cray C90, Cray YMP, NEC machines...). Performance is based on powerfull vector processors. Their memory is physically shared. So access time is uniform and programmation is natural for the modeller.
2. Massively Parallel Processors machines (Paragon, KSR machines, Cray T3D, SP2,CS2...). These machines use a large number of processors and performance is, usually, based on this. For instance, an Intel machine can have several hundreds of micro-processors. So, their cost and the ratio performance over cost are very competitive. Nevertheless, memory is physically distributed among the processors to constitute nodes. These nodes are connected together with a topology : tree, 2D grid, 3D torus, hypercube... So the modeller has to elaborate a strategy for spatial substructuration to optimize communication phases.

We can consider another kind of architecture : a workstation network could be a very interesting solution, for a lab or a SME, to use unemployed computer ressources. Moreover they can take advantages to exploit their own SCAN (Super Computer At Night), instead of spending their limited number of calculation units on Super Computer Centers accounts. Characteristics of workstation network are equivalent to MPP machines ones. But their environment and network performances are less efficient than those of machines devoted to parallel computing.

Our objective is to exploit potential of today and next generation of parallel machines. So we have decided to consider only MPP machines. The MIMD machine that has been chosen for developing code is a 128-processor INTEL iPSC-860 machine.

INTEL iPSC860 machine

The iPSC860 computer installed at ONERA is a 128-node multiprocessor. In this system each processor node is directly connected to 7 neighbouring nodes according to an hypercube topology. Processor activities are synchronised via message passing communication. The communication bandwith is 2.8 MBytes/s and the start up is 175 ms (Berrendorf and Helin, 1991). Usual performance of one individual processor is between 5 and 10 MFlops (64 bits), each node has 8 Mbytes of memory. The programming languages are Fortran and C with communication library for message passing.

Which programming model

For MPP machines, we can consider three different programming models :
1. The Message Passing programming model. The MPP machine is equivalent to a multi-machine (iPSC860, Paragon...). Each processor has its own local memory. So, the modeller organises local and global information transfers to manage the distributed memory.
2. The Shared Virtual Memory programming model (SVM). Each processor can access the global memory. In general, memory is physically addressable in a global way (Cray T3D, CM5...). Shared Virtual Memory could provide high performance during communication phases and ease of programming. Shared virtual memory programming model ensures portability of algorithms and simplicity of concept for parallelisation. Some machines manage local memory like a cache memory. The KSR one uses soft techniques and new architectural one, called ALLCACHEtm memory, to exploit locality by a hierarchical distributed structure of cache memory (Frank, Burkhardt and Rothnie, 1993).
3. The Data Parallelism programming model. This model is based on the splitting of data tables. But, it seems to be efficient on a very limited number of applications (with very regular data structures...).

Nevertheless, we have to consider that memory is physically distributed. So, only distributed algorithms can be efficient on distributed memory computers. Indeed,

even on a shared virtual memory machine, time performance depends on data locality. If the communication manager system has to organise a large number of data transfers through local memories network during a numerical experiment, time performance on a parallel machine could be catastrophic. So, before precise optimization of overlapping computation with communication, locality analysis is necessary to obtain computing performance of a single node and communication performance between nodes. For instance, parallel codes that have bad ratio communication over computation, present bad scalability. Even on a Shared Virtual Memory machine like KSR, we could remark that scalability of BT benchmark (Block Tridiagonal) is better than SP benchmark one (Scalar Pentadiagonal) because of ratio communication over computation (Breit Shah, 1993).
It is why, in Domain Decomposition Methods field for instance, the modeller can preserve and optimise granularity and locality to obtain performance and scalability. So, to be efficient on a parallel system, the modeller has to analyse his code in shared virtual memory context in a same way as in message passing one. Moreover global point of view (shared virtual memory programming model) involves fastidious modification of Fortran loop like in Cray multitasking or autotasking approach. On the contrary, message passing programming model preserves the initial structure of the code. In the Domain Decomposition approach, on each processor or cluster of processors, the same code is running and performing its local experiment. This is SPMD (Single Process Multiple Data) programmation, on each subdomain the modeller's point of view is identical as the initial scalar monodomain code. These considerations lead us to consider that message passing programming model was coherent to exploit Domain Decomposition Approach on a MIMD machine.
Indeed, in the Domain Decomposition approach, a subdomain is allocated to each processor that solves, in a parallel way, its local problem with a duplicated model. Then, the overall solution is generated from local solutions by a communication phase between the processors. These phases call native communication librairies on MIMD machines for local communications, between neighbor subdomains at their common interface, or global computation, a sum or a max on all the subdomains.... So the code structure is preserved and management of dependences is equivalent to an heterogeneous Dirichlet boundary problem. Moreover, in this native communication librairies, blocking functions force tasks synchronisation. On a shared memory programming environment, data management is more complex because the programmer has to manage the boundary problem. For any explicit finite-difference algorithm, calculations remain local and depend only on the point and its neighbours. The dependence problem is therefore equivalent to a non homogeneous boundary problem. For an implicit algorithm, the entire domain is involved, so that it is necessary to use specific numerical methods (Roux, 1990) to change the dependence problem into an open boundary problem.

Moreover, we can use standard communication library (like PVM) for non homogeneous UNIX machines network or (MPI) for devoted MPP machines. Universality of message passing programming model for a large set of numerical model, especially in CFD field, ensures portability on different systems.

Parallelization and dependency analysis

Parallelization strategy is defined by dependency between grid points, spatial extension of operators and their implementation. So, we have decided to consider the static problem of spatial integration, at a given time step, of the Primitive Equation implemented in OPA7. Nevertheless, helped by general field of Domain Decomposition approach, we have taken care to avoid OPA7 specific solutions.
After having identified different damping and forcing terms, advection operators, turbulent diffusion parametrisation...., the physical and dynamical phenomena that contribute to each diagnostic variable eulerian trend, we have isolated two different numerical phases. First, we have an "explicit phase"on horizontal dimensions to perform tridimensional Baroclinic Flow, then we have a two dimensional implicit equation to determine Barotropic Transport. Parallelization strategy of these phases are obviously different due to dependencies.
Indeed, explicit calculation depends only on point and its very local neighbors. If we consider parallelization of a Fortran instructions set from a very formal point of view (Lichnewsky and Thomasset, 1985), we have to preserve the sequential consistency (Leslie Lamport, 1979). To ensure this consistency, we have to respect formal Berstein relations (1966) between input objects (IN) and output objects (OUT) for a given Fortran instruction (IA) at two different iterations.

So, we have :
IN(IA(j)) and OUT(IA(i)) are independent,
OUT(IA(j)) and OUT(IA(i)) are independent,
IN(IA(i)) and OUT(IA(j)) are independent.

In practice, an explicit scheme is rather sequentially consistent, dependencies are very local and could be solved in an efficient way with synchronisation phases like in multitasking Cray or with communication phases in Domain Decomposition approach. We have just to analyse local operators extension and implementation analysis to choose how to oganise data transfers and how to preserve granularity.
In the case of implicit scheme, we have to consider a global operator that involves coupling between all the points in the domain. So there is no locality and we cannot exploit an intrinsic parallelism. We have to implement a new solver to optimize numerical information transfers to perform Barotropic Transport at each time step.

Data sub-structuration

In Domain Decomposition field, the modeller has to choose between two strategies to organise communication at the interface between subdomains.
First, overlapping boundary domains strategy (figure 1) takes advantages of its simplicity.

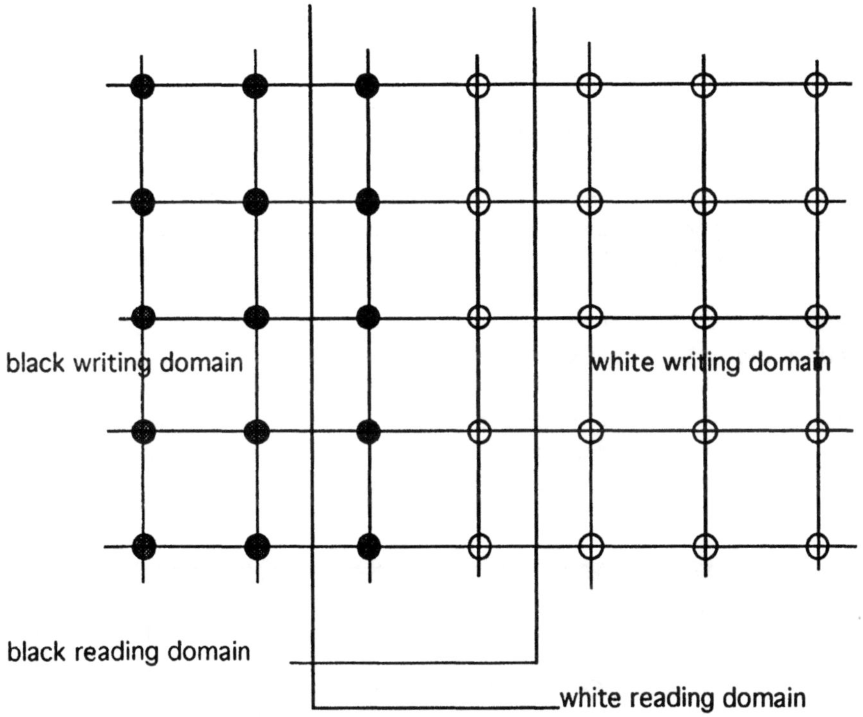

figure 1 an overlapping sub-structuration

The black domain is managed by the "black" processor as the white domain is managed by the "white" processor. Each processor computes on a reading domain. This reading domain owns inner points (the writing domain) plus overlapping interface points (they are also inner points for neighboring processors).
During each time step, overlapping area stocks boundary conditions for computation on writing domain. When the iteration is finished, communication phases start. Computed data of the black writing domain (1) initialize interface data of the white reading domain (2) and vice versa. During such a communication phase, the interface matching conditions are related to open Dirichlet boundary conditions.

X1interfaceread = X2computed,
X2interfaceread = X1computed.

Last but not the least, non overlapping domains strategy presents identical writing and reading domains. This domain owns inner points and "half points" on communication interface and defines a "half scheme" (figure 2).

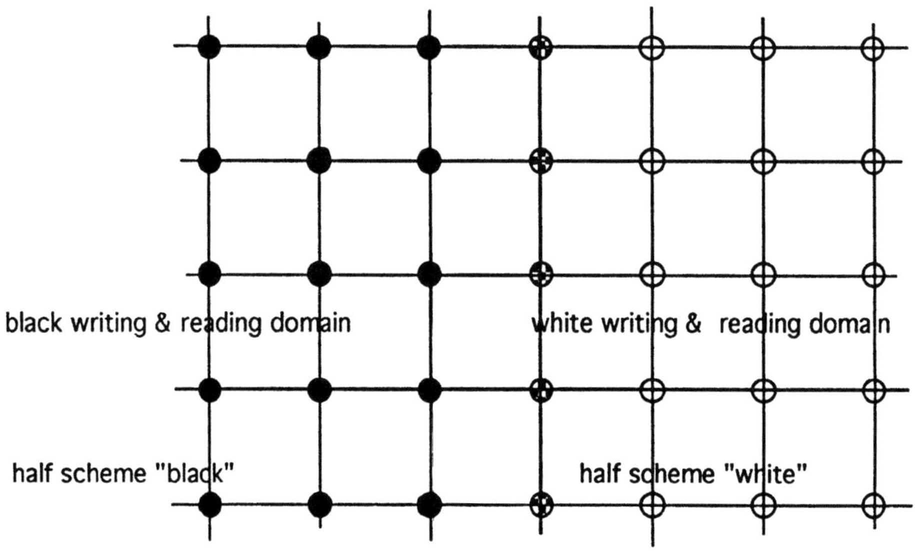

figure 2 a non overlapping sub-structuration

When the time step is finished, a global solution is generated by adding computed local contributions of each neighbors "half schemes".

Xinterface = X1interface "+" X2interface.

To obtain this result, interface explicit operators have to be written as the sum of local neighbors subdomains operators :

Ainterface = A1interface "+" A2interface.

This distribution is, most of the time, natural. If we consider centered difference operator, we can use the following algebraic distribution :

D1interface (X(i)) = (X(i) - X(i-1)),
D2interface (X(i)) = (X(i+1) - X(i)).

Non overlapping strategy presents minimal interface size, so data transfers are optimized as well as memory capacity.
Selecting a data sub-structuration is equivalent to choosing a strategy to solve the open boundary problem. Such a selection should be based on analysis of the local extension and on the implementation of the explicit operators of the code.
The OPA7 finite-difference algorithms are solved using an Arakawa-C grid (figure 3) to reduce numerical noise (Arakawa, 1966). In these algorithms, the local operator are symmetrical, but the Fortran expression becomes non-symmetric.

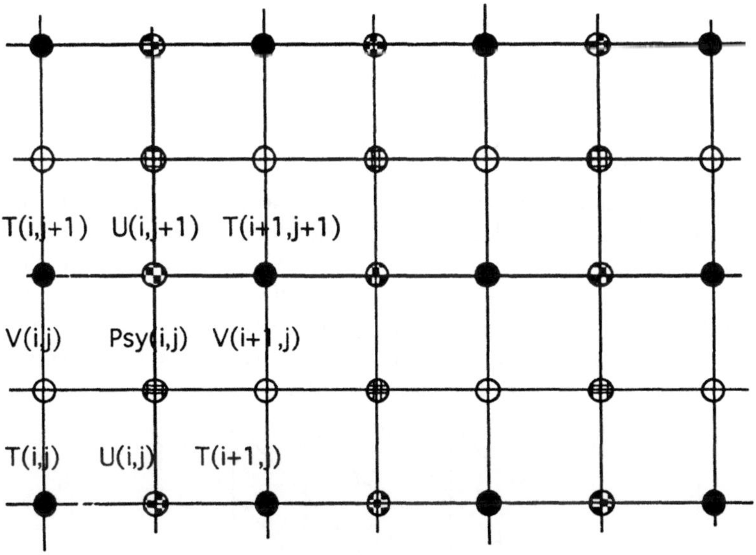

figure 3 an Arakawa-C grid

LEGEND
Upoints and Vpoints are used to compute "honrizontal" velocity
Tpoints are used to compute salinity and temperature.
Psypoints are used to compute Coriolis stress, streamfunction...

Moreover, the OPA7 code has been structured in a large number of subroutines which use local intermediate variables and different splitting of its local operators. These features naturally lead to the easiest programming solution : a data substructuration with overlapping boundaries.(figure 1). In fact, for a complex set of equations, like Primitive Equations of Oceanography, non overlapping strategy would be difficult in practice.
We can notice that strategy choice is imposed by only explicit operators analysis. For parallel solution of Barotropic Transport (implicit phase), we will choose a non overlapping strategy. These different choices will involve just a little more complex implementation of interface management.

Subdomain topology

For finite-difference algorithms, three classical shapes are currently used :
1. Slice splitting as performed in the initial CRAY-multitasking version of the OPA7 code (a one-dimensional topology).
2.Pencil splitting (a two-dimensional topology).
3.Cubic splitting (a three-dimensional topology).

Slice splitting is obviously the easier, this one has been used in other computer developments in Earth Sciences field (Clare and Stevens, 1993). Moreover slice splitting already exists in OPA7 code to use mutitasking Cray technique (Andrich, Madec l'Hostis, 1988).
But, we can notice that, on distributed memory machines, the efficiency of a parallelized code depends on the relative importance of the communication phases with respect to the computation phases. In first approximation, communication phases are assumed to be function of the number of interface points and the computation phases are related to the number of inner points. We can consider a cubic domain with side of n points, so this ratio (granularity) is, at first order, equivalent to n/(2p), wher p is the number of processors (subdomains).

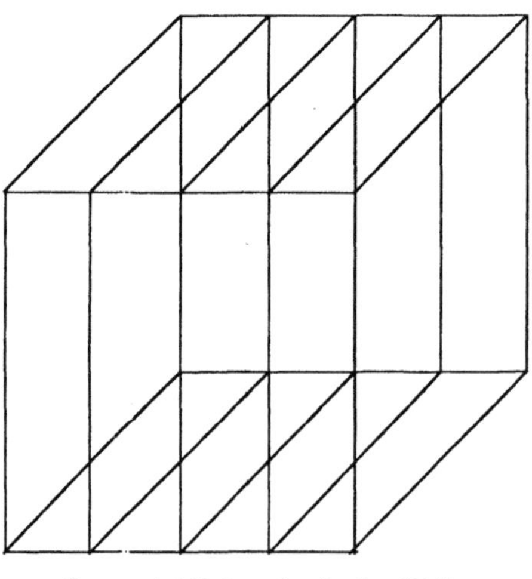

figure 4 1D topological splitting

We can notice that granularity is very weak for a large number of processors. This strategy doesn't ensure scalability in massively parallel computing field. Moreover this one doesn't afford to use the whole interconnecting network.
In fact, optimal splitting strategy to use the whole interconnecting network and to maximize granularity is, at first glance, a very well known mathematical problem. For a given volume (i.e. the whole inner points), we have to find an object that presents a "minimal" surface (i.e. the whole interface). Convexity theory give us a general solution : sphere. If now, we consider the candidates in Finite Differences field, cubic object (figure 5) is a very good candidate and we can verify its granularity to be equivalent to $n/(6\ \sqrt[3]{p})$.

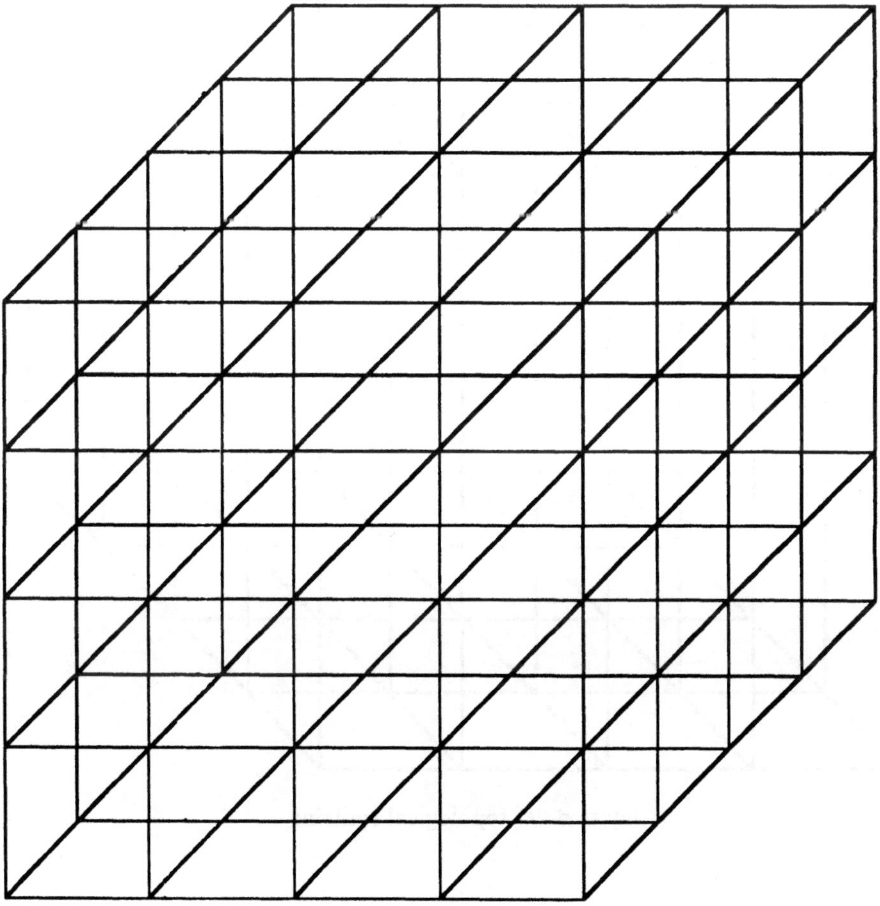

figure 5 3D topological splitting

Nevertheless, ocean fluid dynamics is, at first order, quasi two-dimensional with geostrophic balance on "horizontal" dimensions. This specificity induces specific algorithms on vertical axis (specific mixing, static instability algorithm, surface flux penetration,...), so pencil splitting (figure 6) therefore appears to be adapted to a large set of problems in Earth Sciences field. Moreover, this strategy presents the advantage to be compatible with Barotropic Transport equation arallelization. For this 2D horizontal phase, optimal splitting is a 2D grid.

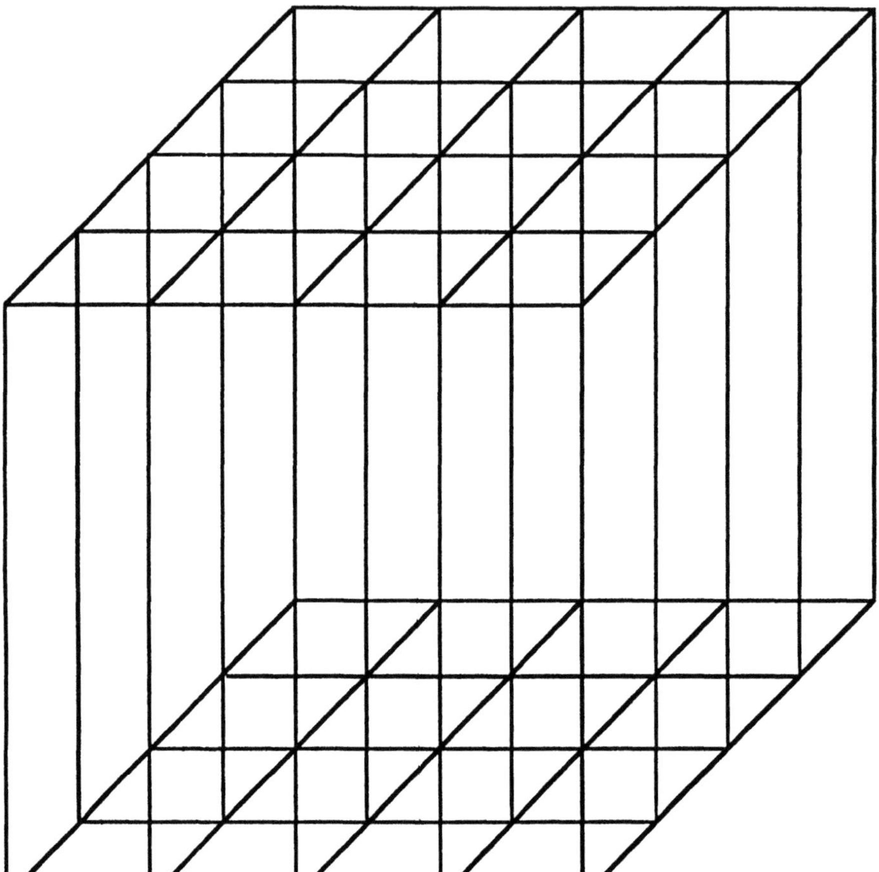

figure 6 2D topological splitting

We can notice this topology has a better ratio "volume" on "surface" (i.e. granularity) than slice splitting (CRAY multitasking). So we consider that pencil strategy is, in Earth Sciences field, a good compromise between massively parallel computer performance and physical specificities of the studied processes. The 2D topological splitting optimizes granularity versus adaptation of such a physical model (OGCM). We can, then preserve conceptual point of view and scalabity for large experiments. This scalability will be possible with finest discretization to observe small scale phenomena. Moreover, we will use more expensive modelisations of physical processes. Conceptual and computer approach lead both to a better understanding of oceanic circulation.

A Domain Decomposition solver

In most of OGCM, the ocean is assumed to be incompressible and its surface is assimilated to a rigid lid to filter fast Barotropic waves. From a physical point of view, streamfunction has been introduce to describe the conservation of velocity vertical integral. Its gradient gi ves water masses transport vertically integrated from the bottom to the surface (Madec, Rahier and Chartier, 1988).
From a mathematical point of view, the streamfunction verifies a partial elliptic differential equation :

div(a(x,y) * gradY) = b(x,y,t).

The matrix of such an operator has "good mathematical propriertics". The matrix is definite, symmetric and positive, so powerfull numerical methods, based on mathematical theory, have been developed. The most standard Domain Decomposition Methods to solve elliptic equation in static field are : the Schwarz Alternative Procedure, the Schur Complement Method and the FETI Method (Finite Elements Tearing and Interconnecting). The FETI Method or Hybrid Method could be interpreted like the dual method of the Schur Complement Method.
The Schwarz Alternative Procedure is an overlapping method and presents no intrinsic parallelism. The FETI Method seems to present better preconditionning proprieties for our kind of problems (Roux, 1989). Moreover, FETI Method is a non overlapping and a non conforming method that enables us to implement zooming procedure. This zooming procedure could provide us a better resolution to perfom local phenomena of small scale for a reasonable computing cost. A well knowledge of physical characteristics will help us to choose interesting area.
So, to compute Barotropic Transport, we have decided to implement a FETI solver. Last, we would like to insist on the point that the FETI approach is a very general approach to solve any elliptic equation.

Conclusion

To conclude, we would like to insist on the very general field of our approach. Domain Decomposition approach is very well suited for parallel computing and message programming model is sufficiently "universal" to perform a large class of

CFD problems. Moreover Domain Decomposition approach and message passing programming model seem to ensure portability on a large number of parallel systems.
In Earth Sciences field, our splitting strategy (pencil topology) respects physical proprierties (geostrophic balance...) and preserves granularity to obtain scalability in massively parallel calculation field. Parallel efficiency is necessary to adopt more sophisticated parametrizations and to use finer discretization.
We have decided to adopt a very easy interface manadgement strategy (with overlapping boundary) to perform Baroclinic Flow and an optimized one (with non overlapping boundary) to solve Barotropic Transport.
Parallelisation using a Domain Decomposition Approach has been applied to the baroclinic (explicit) part of the three-dimensionnal primitive-equation finite-difference OGCM developed in France by LODYC (Guyon, Roux, Chartier and Fraunie, 1994). A pencil topology and an overlapping data substructuration has been fitted to the initial code.
Efficiency tests show a good result with 66 percent efficiency, and a total scalability. The initial structure of the code has been largely preserved (in order to profit by the past debugging achievements). The good "load balancing" obtained between the subdomains illustrates the implementation quality.
Comparisons on a "realistic" numerical test show that a 128-processor iPSC-860 can be better than a 2-processors CRAY-YMP for the OPA-7 code.
These results are very promising in the context of parallel supercomputer development and have confirmed our scientific choices. Further work will consist in the application of the method to the implicit (here barotropic) part of the code.

*Acknowledgements

iPSC-860 and CRAY-YMP are by the courtesy of Office National d'Etude et Recherche Aerospatiale (ONERA). We thank C. Mulard, C. Roge and J. Ryan from ONERA/DI and M. Crepon, P. Delecluseand the Modelling Team from LODYC. Funding has been provided by the Direction des Recherches Etudes et Techniques (numero 92/356) and the frenh Association Nationale de la Recherche Technique (numero 204/92).

*References

ADAM, J.-C. and ALESSANDRINI, V. : Parallelisme et Calcul Intensif de Haute Performance, rapport de l'IDRIS, Institut du Developpement et des Ressources en Informatique Scientifique, Orsay, France, May 1994.

ANDRICH, P., MADEC, G. and L'HOSTIS, D., 1988a : Performance Evaluation for an Ocean General Circulation Model : Vectorization and Multitasking, Conf. Proc Inter. Conf on Supercomputing, July 4-8,St-Malo, France, AMC press, 295-303.

BERRENDORF, R. and HELIN, J. : Analyzing the Performance of Message Passing MIMD Hypercubes: A Study with Intel iPSC/860, International Conference on Supercomputing, June 1991, Cologne, Germany.

BREIT, S. R. and SHAH, G. : Implementing the NAS Parallel Benchmarks on KSR1, Parallel CFD'93, 10-12 May 1993, Paris, France.

CHARTIER, M., 1986: A Three-Dimensional Primitive Equation Ocean General Circulation Model, Rapport CEA-R-5372, Commissariat l'Energie Atomique, Saclay, 31-61.

CLARE, A. R. and STEVENS, D. P. : Implementing finite difference ocean circulation models on MIMD distributed memory computers, Future Generation Computer Systems 9, 1993, 11-18.

FRANCK, S., BURKHARDT III, H. and Rothnie J., Kendall Square Research Corporation : The KSR1 : Bridging the Gap Between Shared Memory and MPPs, Proceeding of Compcon'93, February 1993, San Fransisco, USA, CA,pps 285-294.

GUYON, M., ROUX, F. -X., CHARTIER, M. and FRAUNIE, P. : A Domain Decomposition Method Applied to the Baroclinic Part of an OGCM on a MIMD machine, OCEAN Modelling, ISSUE 103, May 1994.

HERBAUT, C., MORTIER, L. and CREPON, M.: Sentivity Study of the General Circulation of the Western Mediteranean Sea. Part one : The Response to Density Forcing Through the Straits, in preparation.

LICHNEWSKY, A. and THOMASSET, F. : Techniques de Base sur l'Exploitation Automatique du parallelisme dans les Programmes, Conf. Cours INRIA Calcul Parallele a usage Scientifique, Octobre 1985, St-Cyprien, France.

MADEC, G., CHARTIER, M., DELECLUSE, P. and CREPON M. : A Three-Dimensional Numerical Study of Deep-Water Formation in the Northwestern Mediterranean Sea, Journal ofPhysical Oceanography, Vol 21, No. 9, September 1991.

MADEC, G. and CREPON M. : Thermohaline-Driven Deep Water Formation in the Northwestern Mediterranean Sea, P. C.Chu and J. C. Gascard (Editors), Deep Convection and Deep Water Formation in the Oceans, 1991.

MADEC, G., RAHIER, C and CHARTIER, M.: A Comparison of Two-Dimensional Elliptic Solvers for the Barotropic Streamfunction in a Multilevel OGCM, OCEAN Modelling, ISSUE 78, 1988.

MARTI, O., MADEC, G. and DELECLUSE P. : Comment on "Net Diffusivity in Ocean General Circulation Models With Nonuniform Grids", Journal of Geophysical Reseach, 1992.

McBRYAN, O. A. : Performance of the Shallow Water Equations on the CM-200 and CM-5 Parallel Supercomputer, in Proc. of the fifth ECMWF Workshop on the use of Parallel Processors in Meteorology, November 23-27, 1992, Reading, UK.

MESSINGER and ARAKAWA, 1976: Numerical Methods Used in Atmospheric Models, Global Atmospheric Research Program, Publications Series number 17.

NAVON, I. M. and CAI, Y. : Domain decomposition and parallel processing of a finite element model of the Shallow Water Equations, Computer Methods in Applied Mechancs and Engineering, received 19 June 1992.

ROUX, F. -X., Methode de decomposition de domaine l'aide de multiplicateurs de Lagrange et application la resolution en parallele des equations de l'elasticit lineaire, These de doctorat de Paris 6,1989.

ROUX, F.-X., 1990: Domain Decomposition Methods for Static Problem, La Recherche Aerospatiale, n!1, 37-48.

ROUX, F.-X., 1991: Programmation des super-calculateurs scientifiques vectoriels et paralleles, Note Technique 1991-11, ONERA, Chatillon, France.

WOLTERS, L. : Atmosphere and Ocean Circulation Simulation on Massively Parallel Computers, in Proc of the Seventh International Workshop on the Use of Supercomputers in Theoretical Science, June 18-19, 1992.

WOLTERS, L and CATS, G. : A parallel implementation of the Hirlam model, in Proc. of the fifth ECMWF Workshop on the use of Parallel Processors in Meteorology, November 23-27, 1992, Reading, UK.

PREDICIBILITY OF THE OCEANIC CIRCULATIONS.

Christine BERNIER
Département de Mathématiques
Université de Nancy 1
BP 239
54506 Vandoeuvre-lès-Nancy Cedex
France.

ABSTRACT. In this paper, we study the Navier-Stokes equation in which the unknown functions are the vorticity and the streamfunction. Our goal is to estimate the long-time influence of a small error in the initial data. In order to do this, we write down the symmetrized linearized Navier-Stokes equation and we estimate the positive eigenvalues of the corresponding operator.

1. Introduction.

Connected with the problem of data assimilation, the problem of predicibility is now under investigation by many teams [1], [2], [6], [8], [13]. We are interested in the long-time influence of a small error in the initial data, so we linearize and symmetrize the Navier-Stokes equations and study the corresponding linear operator. The local behaviour of a small perturbation is described by the local Lyapounov exponents. They are the eigenvalues of the symmetrized operator. Our aim is to present the results for the multilayer modelisation, but for simplicity, we give the detailed proofs only for a one-layer modelisation, which contains all the difficulties. In this paper, we first recall the well-known results of existence, uniqueness and existence of the attractor for the equation under consideration [4], [10], [15], [16]. Then we study the symmetrized linearized operator. The technical and lengthy part consists in the proof of the continuity in time of its eigenvalues. Finally we establish a priori estimates for the eigenvalues using their variational characterization [12]. In the last part of the paper, we consider the multilayer problem [7], [14], [4], and give a priori estimates of the eigenvalues taking into account the inverse of the internal radii of deformation of each layer.

F.-X. Le Dimet (ed.), High Performance Computing in the Geosciences, 141–158.

2. Existence and Uniqueness Theorems.

Let Ω be a bounded open subset of $\mathbb{R}^2$ with smooth boundary $\partial\Omega$. We consider the system :

$$\begin{cases} \frac{\partial \xi}{\partial t} - \mu\Delta\xi + \frac{\partial \psi}{\partial x} + J(\psi, \xi) = v & \text{in } \Omega\times]0,T[, \\ \xi = 0 & \text{on } \partial\Omega\times]0,T[, \\ \Delta\psi = \xi & \text{in } \Omega\times]0,T[, \\ \psi = 0 & \text{on } \partial\Omega\times]0,T[, \\ \xi(0,x) = \xi_0(x), \end{cases} \tag{1}$$

where $J(f,g) = \frac{\partial f}{\partial x} \times \frac{\partial g}{\partial y} - \frac{\partial f}{\partial y} \times \frac{\partial g}{\partial x}$. We suppose v given in $L^2(\Omega)$. We recall the following results [15], [16], [10], [4].

Proposition 1. For ξ_0 in $H^{-1}(\Omega)$, system (1) admits a unique solution $\xi(t,x)$ in $C([0,T], H^{-1}(\Omega)) \cap L^2(0,T,L^2(\Omega)) \cap L^2_{loc}(0,T,H^1_0(\Omega))$. The semi-group G(t) from $H^{-1}(\Omega)$ in $H^{-1}(\Omega)$, $G(t)\xi_0(x) = \xi(t,x)$, associated to these equations, is such that there exists a maximal attractor A which is bounded in $L^2(\Omega)$, compact and connected in $H^{-1}(\Omega)$ and whose basin of attraction is the whole space $H^{-1}(\Omega)$. This attractor has finite Hausdorff and fractal dimensions. For ξ_0 in $H^1_0(\Omega)$, system (1) admits a unique solution $\xi(t,x)$ in $C([0,T], H^1_0(\Omega)) \cap L^2(0,T,H^2(\Omega))$.

We linearize the system (1) around a solution $\xi(t,x)$ to obtain

$$\begin{cases} \frac{\partial U}{\partial t} - \mu\Delta U + \frac{\partial V}{\partial x} + J(\psi, U) + J(V,\xi) = 0 & \text{in } \Omega\times]0,T[, \\ U = 0 & \text{on } \partial\Omega\times]0,T[, \\ \Delta V = U & \text{in } \Omega\times]0,T[, \\ V = 0 & \text{on } \partial\Omega\times]0,T[, \\ U(0,x) = U_0(x), \end{cases} \tag{2}$$

The existence and the uniqueness of the solution are stated next and follow from general results in [16]:

Proposition 2. For U_0 in $H^{-1}(\Omega)$, system (2) admits a unique solution $U(t,x)$ in $C([0,T], H^{-1}(\Omega)) \cap L^2(0,T,L^2(\Omega))$.

The third proposition links the two systems (1) and (2) (see [16], [10],

[4]).

Proposition 3. For $t > 0$, the semi-group G(t) is uniformly differentiable on A. Its differential in ξ_0 is the linear operator on $H^{-1}(\Omega)$ given by $U \rightarrow L(t_0, \xi_0)U_0 = U(t_0)$, where $U(t_0)$ is the value at time $t = t_0$ of the solution $U(t)$ of the linearized system (2). Moreover,

$$\sup_{\xi_0 \in A} |L(t_0, \xi_0)|_{L(H^{-1})} < \infty.$$

From now on, we consider ξ_0 in $H_0^1(\Omega)$ and we denote by $B = B(t, \xi_0)$ the unbounded operator on $H^{-1}(\Omega)$ defined by

$$BU = -\mu \Delta U + \frac{\partial V}{\partial x} + J(\psi, U) + J(V, \xi), \text{ and } \Delta V = U$$

where $D(B) = H_0^1(\Omega)$. We denote by $S = S(t, \xi_0)$ the symmetric operator $S = \frac{B + B^*}{2}$. We identify $H^{-1}(\Omega)$ with its dual and we introduce on $H^{-1}(\Omega)$ the scalar product $((.,.))$ defined by

$$((u_1, u_2)) = - < u_1, v_2 >_{-1,1} = (\nabla v_1, \nabla v_2)$$

where $(.,.)$ is the usual scalar product in L^2 and where $< .,. >_{-1,1}$ denotes the usual duality pairing between $H^{-1}(\Omega)$ and $H^1(\Omega)$; v_i is defined by $\Delta v_i = u_i$, $v_i = 0$ on $\partial\Omega$. The above scalar product on $H^{-1}(\Omega)$ defines the usual norm $||.||$. We denote by $|.|$ the usual L^2-norm. Let us determine B^*. We recall that an integration by parts shows that $(J(f,g),h) = -(J(f,h),g)$ if one of the three functions f, g, h is equal to zero on the boundary [15], [16], [4]. We then obtain

$$\begin{aligned}
((Bu_1, u_2)) &= - < Bu_1, v_2 >_{-1,1} \\
&= < \mu \Delta u_1, v_2 >_{-1,1} - < \frac{\partial v_1}{\partial x}, v_2 >_{-1,1} - \\
&\quad - < J(\psi, u_1), v_2 >_{-1,1} - < J(v_1, \xi), v_2 >_{-1,1} \\
&= \mu\ (u_1, u_2) + (v_1, \frac{\partial v_2}{\partial x}) + \ (J(\psi, v_2), u_1) + \ (J(v_2, \xi), v_1) \\
&= \mu\ (u_2, u_1) + (\frac{\partial v_2}{\partial x}, v_1) + \ (\Delta J(\psi, v_2), v_1) + \ (J(v_2, \xi), v_1) \\
&= - < -\mu \Delta u_2 - \frac{\partial v_2}{\partial x} - \Delta J(\psi, v_2) - J(v_2, \xi),\ v_1 >_{-1,1} \\
&= ((B^* u_2, u_1)) \\
&= ((u_1, B^* u_2))
\end{aligned}$$

so that

$$B^* u_2 = -\mu \Delta u_2 - \frac{\partial v_2}{\partial x} - \Delta J(\psi, v_2) - J(v_2, \xi).$$

Consequently

$$Su = -\mu \Delta u + \frac{1}{2}[J(\psi, u) - \Delta J(\psi, v)].$$

3. The operator S.

Theorem 1. The operator S is a self-adjoint and closed operator on $H^{-1}(\Omega)$ with compact resolvent. Its eigenvalues λ_i are real and bounded from below by

$$\lambda_i \geq \mu\mu_1 - ||\psi||_{H^2}\sqrt{\mu_1}$$

where μ_1 is the first eigenvalue of the Laplacian operator with homogeneous Dirichlet boundary conditions.
Proof. Let us bound the norm of Su_1 from below. Using successively an integration by parts, the Sobolev's embedding theorem and the Gagliardo-Nirenberg's inequality, we obtain (see [4])

$$|(J(f, v_1), u_1)| \leq ||f||_{H^2(\Omega)} \times |\nabla v_1| \times |u_1|.$$

We now evaluate

$$\begin{aligned} ((Su, u)) &= \mu(\Delta u, v) - \frac{1}{2}[(J(\psi, u), v) - (\Delta J(\psi, v), v)] \\ &\geq \mu(u, u) - ||\psi||_{H^2}\ |\nabla v|\ |u| \\ ((Su, u)) &\geq \mu\ |u|^2 - ||\psi||_{H^2}||u||\ |u|. \end{aligned}$$

So let us consider the operator $T = S + hI$, where I is the identity operator and h a real number such that $h \geq \frac{||\psi||^2_{H^2}}{2\mu}$. We obtain

$$\begin{aligned} ((Tu, u)) &= ((Su, u)) + h((u, u)) \\ ((Tu, u)) &\geq \mu\ |u|^2 - ||\psi||_{H^2}||u||\ |u| + h||u||^2 \\ ((Tu, u)) &\geq \left(\sqrt{\frac{\mu}{2}}|u| - \frac{||\psi||_{H^2}}{\sqrt{2\mu}}||u||\right)^2 + \frac{\mu}{2}|u|^2 \\ ||Tu||\ ||u|| &\geq \frac{\mu}{2}|u|^2 \\ ||Tu||\ ||u|| &\geq \frac{\mu\sqrt{\mu_1}}{2}|u|\ ||u|| \end{aligned}$$

$$||Tu|| \geq \frac{\mu\sqrt{\mu_1}}{2}|u|.$$

Therefore the operator T has a bounded inverse. Then T is a closed operator and consequently so is S. Now, using the compact embedding from $L^2(\Omega)$ into $H^{-1}(\Omega)$, we prove that T is an operator with compact resolvent. We then conclude that the same property holds true for S. The operator S is a self-adjoint operator with compact resolvent. Its eigenvalues are real. We now bound them from below. We have already seen that

$$((Su,u)) + h((u,u)) \geq \mu\ |u|^2 - ||\psi||_{H^2}||u||\ |u| + h||u||^2.$$

Let ϵ be a positive real number. We have for $h \geq \frac{||\psi||^2_{H^2}}{4\epsilon\mu}$ the estimate

$$((Su,u)) + h((u,u)) \geq \left(\sqrt{\epsilon\mu}|u| - \frac{||\psi||_{H^2}}{2\sqrt{\epsilon\mu}}||u||\right)^2 + (1-\epsilon)\mu|u|^2$$

so that if u is an eigenvector corresponding to the eigenvalue λ, we obtain

$$\begin{aligned} (\lambda + h)||u||^2 &\geq (1-\epsilon)\mu\mu_1||u||^2 \\ (\lambda + h) &\geq (1-\epsilon)\mu\mu_1 \\ \lambda &\geq (1-\epsilon)\mu\mu_1 - h. \end{aligned}$$

In particular, for $\epsilon = \frac{1}{2}$ we obtain :

$$\lambda \geq \frac{\mu\mu_1}{2} - \frac{||\psi||^2_{H^2}}{2\mu}. \tag{3}$$

Optimizing in ϵ, we get

$$\lambda \geq \mu\mu_1 - ||\psi||\sqrt{\mu_1}.$$

From now on, the eigenvalues are ordered in such a way that : $\lambda_1 \leq \lambda_2 \leq$ ...

4. Eigenvalues of the operator S and predicibility.

Let us consider m solutions of system (2), $U_1(t), U_2(t), .., U_m(t)$, corresponding to the m initial data $U_1^0, U_2^0, ..., U_m^0$. We denote by $|\ U_1^0 \ \wedge\ U_2^0\ \wedge\ ...\ \wedge\ U_m^0\ |$ the m-volume of the parallelepiped with edges $U_1^0, U_2^0, ..., U_m^0$, and by $|\ U_1(t)\ \wedge\ U_2(t)\ \wedge\ ...\ \wedge\ U_m(t)\ |$ the m-volume of the parallelepiped with edges $U_1(t), U_2(t), .., U_m(t)$. Define

$$\omega_m(t,\xi_0) = \sup_{\substack{U_i^0 \in H^{-1}(\Omega) \\ ||U_i^0|| \leq 1}} \frac{|\ U_1(t)\ \wedge\ U_2(t)\ \wedge\ ...\ \wedge\ U_m(t)\ |}{|\ U_1^0\ \wedge\ U_2^0\ \wedge\ ...\ \wedge\ U_m^0\ |}.$$

The numbers $\omega_m(t,\xi_0)$ determine the largest distortion of an infinitesimal m-dimensional volume generated by $G(t)$ around the point $\xi_0 \in H^{-1}(\Omega)$. One can prove that [16]

$$| U_1(t) \wedge ... \wedge U_m(t) | = | U_1^0 \wedge ... \wedge U_m^0 | \, exp\left(\int_0^t Tr(-B(s,\xi_0)oQ_m(s))ds\right)$$

where $Q_m(s) = Q_m(s, U_1^0, U_2^0, ..., U_m^0)$ is the projector from $H^{-1}(\Omega)$ onto the space spanned by $U_1(t), U_2(t), .., U_m(t)$, and where Tr denotes the trace operator. It follows that

$$\omega_m(t,\xi_0) = \sup_{\substack{U_i^0 \in H^{-1}(\Omega) \\ \|U_i^0\| \leq 1}} exp\left(\int_0^t Tr(-B(s,\xi_0)oQ_m(s))ds\right)$$

and so

$$\frac{1}{t} log\omega_m(t,\xi_0) = \sup_{\substack{U_i^0 \in H^{-1}(\Omega) \\ \|U_i^0\| \leq 1}} \left(\frac{1}{t}\int_0^t Tr(-B(s,\xi_0)oQ_m(s))ds\right) \tag{4}$$

But by definition, we have

$$Tr(-B(s,\xi_0)oQ_m(s)) = \sum_{i=1}^{i=m}((-B\phi_i(s), \phi_i(s)\ ))$$

where $\phi_i(s), i \in \mathbb{N}$, is an orthonormal basis of $H^{-1}(\Omega)$, $\phi_i(s) \in L^2(\Omega)$, with $\phi_1(s), ..., \phi_m(s)$ spanning $Q_m(s)H^{-1}(\Omega) = Span\,[U_1(t), U_2(t), .., U_m(t)]$. Noticing that

$$((-B\phi_i(s), \phi_i(s)\ )) = ((-S\phi_i(s), \phi_i(s)\ )),$$

we conclude that

$$Tr(-B(s,\xi_0)oQ_m(s)) = Tr(-S(s,\xi_0)oQ_m(s)),$$

and, since S is self-adjoint,

$$Tr(-S(s,\xi_0)oQ_m(s)) = \sum_{i=1}^{i=m}((-S\phi_i(s), \phi_i(s)\ )) \leq \sum_{i=1}^{i=m} \lambda_i(-S(s,\xi_0)).$$

From inequality (4), we obtain

$$\lim_{t\to 0} \frac{1}{t}\, log\omega_m(t,\xi_0) \ \leq\ \lim_{t\to 0} \left(\frac{1}{t}\int_0^t \sum_{i=1}^{i=m} \lambda_i(-S(s,\xi_0))\, ds\right) \tag{5}$$

We now prove :

Theorem 2. The following equality holds :

$$\lim_{t\to 0} \frac{1}{t}\, log\omega_m(t,\xi_0) = \sum_{i=1}^{i=m} \lambda_i(-S(0,\xi_0)) \tag{6}$$

We first prove the following.
Lemma. The eigenvalue $\lambda(S(t,\xi_0))$ converges to $\lambda(S(0,\xi_0))$ continuously when t tends to zero.
Remark. The continuity in time of the operator $Tr(-S(t,\xi_0)oQ_m(t))$ follows from this lemma.

We divide the proof into 3 steps.

First step. The operator S is continuous in time.

We prove the continuity of S in the H^{-1}-norm. We have

$$\begin{aligned} ||S(t)U - S(t_0)U|| &= \frac{1}{2}||J(\psi(t),U) - \Delta J(\psi(t),V) - \\ &\quad -J(\psi(t_0),U) + \Delta J(\psi(t_0),V)|| \\ &= \frac{1}{2}||J(\psi(t)-\psi(t_0),U) - \Delta J(\psi(t)-\psi(t_0),V)|| \\ &= \frac{1}{2}||AU|| \end{aligned}$$

where $AU = J(\psi(t)-\psi(t_0),U) - \Delta J(\psi(t)-\psi(t_0),V)$.
We introduce $\Psi = \psi(t)-\psi(t_0)$, $\Xi = \xi(t)-\xi(t_0)$. Let us evaluate $||AU||$. We have

$$\begin{aligned} < AU, v > &= < J(\Psi,U), v > - < \Delta J(\Psi,V), v >, \\ < AU, v > &= < J(\Psi,\Delta V), v > - < J(\Psi,V), u > . \end{aligned}$$

An integration by part of the first term leads to the estimates :

$$\begin{aligned} < AU, v > &\leq | < J(\Xi,V), v > | + 2\int |v_x J(\Psi_x,V) + v_y J(\Psi_y,V)|\, d\omega, \\ < AU, v > &\leq |\nabla\Xi||V||_\infty|\nabla v| + 2\left(|v_x||\nabla\Psi_x|_{L^4} + |v_y||\nabla\Psi_y|_{L^4}\right)|\nabla v|_{L^4}, \\ &\leq |\nabla\Xi||\nabla v||U| + c|\nabla v||\nabla\Xi||U|, \\ &\leq c'|\nabla\Xi||\nabla v||U|. \end{aligned}$$

So $||AU|| \leq c'|U||\nabla\Xi|$ and then

$$||S(t)U - S(t_0)U|| \leq c'|U|\;|\nabla\xi(t) - \nabla\xi(t_0)|.$$

Since $\nabla\xi(t)$ is continuous in time, we obtain the result.

Second step. The operator S is continuous in the "generalized sense" with respect to the viscosity μ.

Proof. We denote by S_n the closed operator in $H^{-1}(\Omega)$ defined by

$$S_n u_n = -\left(\mu + \frac{1}{n}\right)\Delta u_n + \frac{1}{2}\left(J(\psi, u_n) - \Delta J(\psi, v_n)\right)$$

and prove that S_n converges to S in the generalized sense defined by Kato [11]. Following [11], let us recall a useful criterion for convergence "in the generalized sense". Let T be a closed operator in H, with a non-empty resolvent set $\rho(T)$. We denote by $R(\zeta, T)$ the resolvent of T.
Lemma. In order that a sequence T_n of closed operators converges to T in the generalized sense, it is necessary that each $\zeta \in \rho(T)$ belong to $\rho(T_n)$ for sufficiently large n and that $||R(\zeta, T_n) - R(\zeta, T)|| \to 0$ when $n \to \infty$, while it is sufficient that this be true for some $\zeta \in \rho(T)$.

We know that $\lambda < \frac{\mu\mu_1}{2} - \frac{||\psi||^2_{H^2}}{2\mu}$ implies $\lambda \in \rho(T)$. So we now consider $\zeta \geq \frac{||\psi||^2_{H^2}}{2\mu}$ and estimate $||R(-\zeta, T_n) - R(-\zeta, T)||$. We introduce u_n and u the solutions of $(S_n + \zeta I)u_n = f$ and $(S + \zeta I)u = f$. Substracting these two equations, we obtain

$$S_n u_n - Su + \zeta(u_n - u) = 0.$$

Since $S_n u_n = Su_n - \frac{1}{n}\Delta u_n$, it follows that $S(u_n - u) + \zeta(u_n - u) = \frac{1}{n}\Delta u_n$. We write $w = u_n - u$ and $z = v_n - v$ and we recall that $\Delta z = w$. We then rewrite the last equation as

$$-\mu\Delta w + \frac{1}{2}\left(J(\psi, w) - \Delta J(\psi, z)\right) + \zeta w = \frac{1}{n}\Delta u_n.$$

Multiplying by z and integrating on Ω, we obtain:

$$\begin{aligned}
\mu|w|^2 - (J(\psi, w), z) + \zeta|\nabla z|^2 &\leq \frac{1}{n}|u_n||w|, \\
\mu|w|^2 + \zeta|\nabla z|^2 &\leq \frac{1}{n}|u_n||w| + ||\psi||_{H^2}|\nabla z||w|, \\
&\leq \mu|w|^2 + \frac{1}{2n^2\mu}|u_n|^2 + \frac{1}{2\mu}||\psi||^2_{H^2}|\nabla z|^2
\end{aligned}$$

and so

$$|\nabla z|^2\left(\zeta - \frac{1}{2\mu}||\psi||^2_{H^2}\right) \leq \frac{1}{2n^2\mu}|u_n|^2. \tag{7}$$

We now estimate $|u_n|^2$. Multiplying the equation $(S_n + \zeta I)u_n = f$ by v_n and integrating on Ω, we have

$$\begin{aligned}\left(\mu + \frac{1}{n}\right)|u_n|^2 + \zeta|\nabla v_n|^2 &\leq ||f||\,|\nabla v_n| + |(J(\psi, u_n), v_n)| \\ &\leq ||f||\,|\nabla v_n| + ||\psi||_{H^2}\,|\nabla v_n|\,|u_n| \\ &\leq ||f||\,|\nabla v_n| + \frac{\mu}{2}|u_n|^2 + \frac{||\psi||^2_{H^2}}{2\mu}|\nabla v_n|^2.\end{aligned}$$

Consequently, we get

$$\begin{aligned}\frac{\mu}{2}|u_n|^2 + \left(\zeta - \frac{||\psi||^2_{H^2}}{2\mu}\right)|\nabla v_n|^2 &\leq ||f||\,|\nabla v_n| \\ \frac{\mu}{2}|u_n|^2 + \left(\zeta - \frac{||\psi||^2_{H^2}}{2\mu}\right)|\nabla v_n|^2 &\leq ||f||\,\sqrt{\lambda_1}|u_n| \\ \frac{\mu}{2}|u_n| &\leq ||f||\,\sqrt{\lambda_1} \\ |u_n| &\leq \frac{2\,||f||\,\sqrt{\lambda_1}}{\mu}.\end{aligned}$$

Inserting the last estimate into (7), we obtain

$$\begin{aligned}|\nabla z|^2\left(\zeta - \frac{1}{2\mu}||\psi||^2_{H^2}\right) &\leq \frac{2}{n^2\mu^3}||f||^2\,\lambda_1 \\ \frac{|\nabla z|}{||f||} &\leq \frac{1}{\mu n}\sqrt{\frac{4\lambda_1}{2\mu\zeta - ||\psi||^2_{H^2}}}.\end{aligned}$$

Since

$$\begin{aligned}||R(-\zeta, T_n) - R(-\zeta, T)|| &= \sup_{f\in H^{-1}} \frac{||u_n - u||}{||f||} \\ &= \sup_{f\in H^{-1}} \frac{|\nabla v_n - \nabla v|}{||f||} \\ &= \sup_{f\in H^{-1}} \frac{|\nabla z|}{||f||},\end{aligned}$$

we get

$$||R(-\zeta, T_n) - R(-\zeta, T)|| \leq \frac{1}{\mu n}\sqrt{\frac{4\lambda_1}{2\mu\zeta - ||\psi||^2_{H^2}}} \to 0$$

as n goes to infinity. This proves the convergence in the generalized sense of S_n to S.

Third step. We now prove the lemma.

We consider U such that $|\nabla V| = 1$ and $E_t(U) = ((S(t,\xi_0)U,U))$ and $E_t^n(U) = ((S_n(t,\xi_0)U,U))$. We fix n. By the continuity of $|\nabla\xi|$, there exists a time T such that for all t $\leq$ T, $|\nabla\xi(t) - \nabla\xi(0)| \leq \frac{2}{n}$. By the continuity of S in time, we obtain

$$|E_t(U) - E_0(U)| \leq \frac{2}{n}|U| \leq \frac{1}{n} + \frac{|U|^2}{n}$$

$$E_0(U) - \frac{1}{n} - \frac{|U|^2}{n} \leq E_t(U) \leq E_0(U) + \frac{1}{n} + \frac{|U|^2}{n}$$

But $E_0(U) - \frac{|U|^2}{n} = E_0^{-1/n}(U)$ and $E_0(U) + \frac{|U|^2}{n} = E_0^{1/n}(U)$. Then, taking the inf-sup definition of the eigenvalues, it follows that

$$\lambda_i^{-1/n}(S(0,\xi_0)) - \frac{1}{n} \leq \lambda_i(S(t,\xi_0)) \leq \lambda_i^{1/n}(S(0,\xi_0)) + \frac{1}{n}.$$

We now use a result of [11]. Let T be a closed operator in H.
Lemma. If a sequence T_n of closed operators converges to T in the generalized sense, then a finite system of eigenvalues of T_n converges continuously to the corresponding finite system of eigenvalues of T.
Since S_n converges to S in the generalized sense, the eigenvalues are continuous in n. So when t tends to 0, n tends to infinity and $\lambda_i^{1/n}(S(0,\xi_0))$ and $\lambda_i^{-1/n}(S(0,\xi_0))$ tend to $\lambda_i(S(0,\xi_0))$. Consequently $\lambda_i(S(t,\xi_0))$ tends to $\lambda_i(S(0,\xi_0))$.

Proof of theorem 2.

Since the operator $Tr(-S(s,\xi_0)oQ_m(s))$ is continuous in s, we have

$$Tr(-S(0,\xi_0)oQ_m(0)) = \lim_{t\to 0}\frac{1}{t}\int_0^t Tr(-S(s,\xi_0)oQ_m(s))ds.$$

And so by (4), it follows that

$$Tr(-S(0,\xi_0)oQ_m(0)) \leq \lim_{t\to 0}\frac{1}{t}log\omega_m(t,\xi_0).$$

Taking the supremum on the left side, we get

$$\sup_{\substack{U_i^0 \in H^{-1}(\Omega)\\ ||U_i^0|| \leq 1}} Tr(-S(0,\xi_0)oQ_m(0)) \leq \lim_{t\to 0}\frac{1}{t}log\omega_m(t,\xi_0) \tag{8}$$

We now consider an orthonormal basis $\phi_1, ..., \phi_m$ consisting of the first m eigenvectors of $S(0,\xi_0)$ and we denote by Q_o the projector in $H^{-1}(\Omega)$ onto $V_o = Span\,[\phi_1, ..., \phi_m]$. So we have

$$Tr(-S(0,\xi_0)oQ_o) = \sum_{i=1,...,m} ((-S\phi_i, \phi_i)) = \sum_{i=1,...,m} \lambda_i(-S(0,\xi_0)).$$

But we also have

$$Tr(-S(0,\xi_0)oQ_o) \leq \sup_{\substack{U_i^0 \in H^{-1}(\Omega) \\ \|U_i^0\| \leq 1}} Tr(-S(0,\xi_0)oQ_m(0))$$

and consequently

$$\sum_{i=1,...,m} \lambda_i(-S(0,\xi_0))) \leq \sup_{\substack{U_i^0 \in H^{-1}(\Omega) \\ \|U_i^0\| \leq 1}} Tr(-S(0,\xi_0)oQ_m(0)).$$

This result, together with estimate (8), gives us

$$\sum_{i=1,...,m} \lambda_i(-S(0,\xi_0))) \leq \lim_{t\to 0} \frac{1}{t} log\omega_m(t,\xi_0).$$

To obtain the reverse inequality, we recall that (see (5) above)

$$\lim_{t\to 0} \frac{1}{t}\; log\omega_m(t,\xi_0) \;\leq\; \lim_{t\to 0} \left(\frac{1}{t}\int_0^t \sum_{i=1}^{i=m} \lambda_i(-S(s,\xi_0))\; ds\right).$$

Using now the continuity of the eigenvalues of S, the result follows.

5. Estimates of the eigenvalues of S.

We estimate the eigenvalues of S using their variational characterization [12]. Let

$$E(U) = ((SU,U)) = \mu|U|^2 + (J(\psi,V),U).$$

We are looking for the eigenvalues of E under the constraint K :

$$K = \left\{U \in L^2(\Omega), ||U|| = (|\nabla V| =)1\right\}.$$

As before , we bound the Jacobian term from above as

$$|(J(\psi,V),U)| \leq ||\psi||_{H^2}\; |\nabla V|\; |U|.$$

Since $|\nabla V| = 1$, we use Young's inequaliy to get

$$|(J(\psi,V),U)| \leq \mu\epsilon|U| + \frac{1}{4\epsilon\mu}||\psi||_{H^2}^2.$$

So we obtain

$$\mu(1-\epsilon)|U|^2 - \frac{1}{4\epsilon\mu}||\psi||_{H^2}^2 \leq E(U) \leq \mu(1+\epsilon)|U|^2 + \frac{1}{4\epsilon\mu}||\psi||_{H^2}^2 . \qquad (9)$$

We now consider the functional $F(U) = |U|^2$. The critical points μ_k of F under the constraint K are the eigenvalues of Laplace's operator with Dirichlet homogeneous boundary conditions. They are characterized by

$$\mu_k = \inf_{\substack{A \in L^2(\Omega) \\ \dim A = k}} \sup_{\substack{U \in A \\ ||U||=1}} F(U).$$

In the same way, the eigenvalues λ_k of S are given by

$$\lambda_k = \inf_{\substack{A \in L^2(\Omega) \\ \dim A = k}} \sup_{\substack{U \in A \\ ||U||=1}} E(U).$$

So, taking the supremum over $U \in A$ in (9), and then the infimum over all of the subspaces A of dimension k, we obtain for $\epsilon \in]0,1]$:

$$\mu(1-\epsilon)\mu_k - \frac{1}{4\epsilon\mu}||\psi||_{H^2}^2 \leq \lambda_k \leq \mu(1+\epsilon)\mu_k + \frac{1}{4\epsilon\mu}||\psi||_{H^2}^2.$$

We optimize on ϵ and use the estimation (3) to obtain

Theorem 3. The eigenvalues λ_k of S satisfy

$$\lambda_k \geq max(e_1, e_k) \qquad (10)$$

where

$$\begin{aligned} e_1 &= \mu\mu_1 - ||\psi||_{H^2}\sqrt{\mu_1}, \\ e_k &= \mu\mu_k - ||\psi||_{H^2}\sqrt{\mu_k}, \end{aligned}$$

and where μ_k is the kth eigenvalue of Laplace's operator with Dirichlet homogeneous boundary conditions.

6. The multilayer problem.

Here, we are looking for similar a priori estimates of the eigenvalues when we study the multilayer modelization. For simplicity, we consider only two layers with Dirichlet boundary conditions [7], [14], [3]. The system under consideration is the following :

$$\begin{cases} \dfrac{\partial \theta_1}{\partial t} - \mu\Delta^2\psi_1 + J(\psi_1, \theta_1) = \dfrac{v}{H_1} & \text{in} \quad \Omega\times]0,T[, \\ \dfrac{\partial \theta_2}{\partial t} - \mu\Delta^2\psi_2 + J(\psi_2, \theta_2) = 0 & \text{in} \quad \Omega\times]0,T[, \\ \theta = 0 & \text{on} \quad \partial\Omega\times]0,T[, \\ \Delta\psi - W\psi = \theta & \text{in} \quad \Omega\times]0,T[, \\ \psi = 0 & \text{on} \quad \partial\Omega\times]0,T[, \\ \theta(0,x) = \theta_0(x), \end{cases} \tag{11}$$

where θ is the vector with components θ_1, θ_2, ψ the vector with components ψ_1, ψ_2, and W is the 2×2 matrix defined by

$$W = p \begin{pmatrix} \dfrac{1}{H_1} & -\dfrac{1}{H_1} \\ -\dfrac{1}{H_2} & \dfrac{1}{H_2} \end{pmatrix}.$$

The kth layer is characterized by its thickness H_k and its streamfunction ψ_k. The reduced gravity g' and the parameter of Coriolis in the middle of the basin f define the parameter $p = \dfrac{f^2}{g'}$.
The existence and the uniqueness of a solution of system (11) and the existence of a finite dimensional attractor are proved in [4].

We now consider the linearized system :

$$\begin{cases} \dfrac{\partial U_1}{\partial t} - \mu\Delta^2 V_1 + J(\psi_1, U_1) + J(V_1, \theta_1) = 0 & \text{in} \quad \Omega\times]0,T[, \\ \dfrac{\partial U_2}{\partial t} - \mu\Delta^2 V_2 + J(\psi_2, U_2) + J(V_2, \theta_2) = 0 & \text{in} \quad \Omega\times]0,T[, \\ U = 0 & \text{on} \quad \partial\Omega\times]0,T[, \\ \Delta V - WV = U & \text{in} \quad \Omega\times]0,T[, \\ V = 0 & \text{on} \quad \partial\Omega\times]0,T[, \\ U(0,x) = U_0(x). \end{cases} \tag{12}$$

We introduce on $H^{-1}(\Omega)$ the norm defined by the scalar product [4] :

$$((U,u)) = -H_1\ (U_1, v_1) - H_2\ (U_2, v_2).$$

We remark that

$$||U||^2 = H_1\ |\nabla V_1|^2 + H_2\ |\nabla V_2|^2 + p\ |V_1 - V_2|^2.$$

This norm is equivalent to the usual one. The operator B is defined by

$$BU = \begin{pmatrix} -\mu\Delta^2 V_1 + J(\psi_1, U_1) + J(V_1, \theta_1) \\ -\mu\Delta^2 V_2 + J(\psi_2, U_2) + J(V_2, \theta_2) \end{pmatrix}.$$

As before, we introduce $S = \dfrac{B + B^*}{2}$.

Theorem 4. The eigenvalues λ_k of S verify

$$\lambda_k \geq max(e_1, e_k)$$

where

$$\begin{aligned} e_1 &= \mu\mu_1\beta - \sqrt{2\mu_1\beta\left(||\psi_1||^2_{H^2}\left[1 + \frac{p}{H_1}\right] + ||\psi_2||^2_{H^2}\left[1 + \frac{p}{H_2}\right]\right)}, \\ e_k &= \mu\nu_k - \sqrt{2\nu_k\left(||\psi_1||^2_{H^2}\left[1 + \frac{p}{H_1}\right] + ||\psi_2||^2_{H^2}\left[1 + \frac{p}{H_2}\right]\right)}. \end{aligned}$$

and

$$\nu_k \in \left\{ \begin{array}{c} \mu_i, \; i \in \mathbb{N} \\ \dfrac{\mu_j^2}{\mu_j + \left(\frac{p}{H_1} + \frac{p}{H_2}\right)}, j \in \mathbb{N} \end{array} \right.$$

Proof of the theorem. We introduce the functional $E(U)$ defined by

$$E(U) = ((SU, U)) = ((BU, U)).$$

We obtain :

$$\begin{aligned} E(U) &= \mu\,(H_1|\Delta U_1|^2 + H_2|\Delta U_2|^2) - \\ &\quad -H_1(J(V_1, \psi_1), \Delta V_1) - p(J(V_1, \psi_1), (V_1 - V_2)) - \\ &\quad -H_2(J(V_2, \psi_2), \Delta V_2) - p(J(V_2, \psi_2), (-V_1 + V_2)). \end{aligned}$$

Our aim is to find the critical points of $E(U)$ under the constraint K

$$K = \left\{U \in L^2(\Omega), ||U|| = 1\right\}.$$

We denote by $J(U)$ the sum of the terms with the Jacobian operator :

$$\begin{aligned} J(U) &= -H_1(J(V_1, \psi_1), \Delta V_1) - (J(V_1, \psi_1), p(V_1 - V_2)) - \\ &\quad -H_2(J(V_2, \psi_2), \Delta V_2) - (J(V_2, \psi_2), p(-V_1 + V_2)). \end{aligned}$$

Let us now bound $J(U)$ from above. Let $\epsilon \geq 0$; we have

$$\begin{aligned} |H_i(J(V_i, \psi_i), \Delta V_i)| &\leq ||\psi_i||_{H^2}\, |\sqrt{H_i}\nabla V_i|\, |\sqrt{H_i}\Delta V_i|, \\ &\leq \mu H_i \frac{\epsilon}{2}\, |\Delta V_i|^2 + \frac{||\psi_i||^2_{H^2}}{2\mu\epsilon}\, |\nabla V_i|^2 \end{aligned}$$

and

$$
\begin{aligned}
|p(J(V_i,\psi_i),(V_1-V_2))| &\leq \frac{||\psi_i||_{H^2}}{\sqrt{H_i}}\ p\ |V_1-V_2|\ |\sqrt{H_i}\Delta V_i| \\
&\leq \mu H_i \frac{\epsilon}{2}\ |\Delta V_i|^2 + \frac{||\psi_i||^2_{H^2}\ p^2}{2\mu\epsilon\ H_i}\ |V_1-V_2|^2.
\end{aligned}
$$

So

$$
\begin{aligned}
|J(U)| \leq\ & \mu\epsilon \left[H_1|\Delta V_1|^2 + H_2|\Delta V_2|^2\right] + \\
& + \frac{1}{2\mu\epsilon}\left(H_1\ ||\psi_1||^2_{H^2}|\nabla V_1|^2 + H_2\ ||\psi_2||^2_{H^2}|\nabla V_2|^2\right) + \\
& + \frac{1}{2\mu\epsilon}\left(||\psi_1||^2_{H^2}\frac{p}{H_1} + ||\psi_2||^2_{H^2}\frac{p}{H_2}\right)\ p\ |V_1-V_2|^2.
\end{aligned}
$$

But under the constraint K, we have $p\ |V_1-V_2|^2 \leq 1$ and $H_i|\nabla V_i|^2 \leq 1$. Therefore

$$
|J(U) \leq \mu\epsilon\left(H_1|\Delta V_1|^2 + H_2|\Delta V_2|^2\right) + \tag{13}
$$

$$
+\frac{1}{2\mu\epsilon}\left(||\psi_1||^2_{H^2}\left[1+\frac{p}{H_1}\right] + ||\psi_2||^2_{H^2}\left[1+\frac{p}{H_2}\right]\right). \tag{14}
$$

We can now estimate $E(U)$ as follows :

$$
\begin{aligned}
E(U) &\geq \mu\left(H_1|\Delta V_1|^2 + H_2|\Delta V_2|^2\right) - |J(U)| \\
E(U) &\geq \mu(1-\epsilon)\left(H_1|\Delta V_1|^2 + H_2|\Delta V_2|^2\right) - \\
& -\frac{1}{2\mu\epsilon}\left(||\psi_1||^2_{H^2}\left[1+\frac{p}{H_1}\right] + ||\psi_2||^2_{H^2}\left[1+\frac{p}{H_2}\right]\right).
\end{aligned}
$$

We first look for an a priori estimate for all the eigenvalues. Let us denote by R the second factor of the last inequality. We consider $((SU,U))+R((U,U))$. We obtain :

$$
\begin{aligned}
((SU,U)) + R((U,U)) &= E(U) + R||U||^2 \\
((SU,U)) + R((U,U)) &\geq \mu(1-\epsilon)\left(H_1|\Delta V_1|^2 + H_2|\Delta V_2|^2\right) + \\
&\quad + R\ (||U||^2 - 1) \\
((SU,U)) + R((U,U)) &\geq \mu_1\mu(1-\epsilon)\left(H_1|\nabla V_1|^2 + H_2|\nabla V_2|^2\right) + \\
&\quad + R\ (||U||^2 - 1).
\end{aligned}
$$

We consider U an eigenvector of S with eigenvalue λ such that $||U|| = 1$; then

$$
(\lambda + R) \geq \mu(1-\epsilon)\mu_1\left(H_1|\nabla V_1|^2 + H_2|\nabla V_2|^2\right).
$$

From the inequality

$$\frac{H_1|\nabla V_1|^2 + H_2|\nabla V_2|^2}{2} \geq min(H_1, H_2)\frac{\sqrt{\mu_1}}{4}|V_1 - V_2|^2,$$

it follows that

$$(\lambda + R) \geq \mu(1-\epsilon)\mu_1 min(\frac{1}{2}, \frac{min(H_1, H_2)\sqrt{\mu_1}}{4p}.$$

Then, introducing $\beta = min\left(\frac{1}{2}, \frac{min(H_1, H_2)\sqrt{\mu_1}}{4p}\right)$, we obtain

$$\lambda \geq \mu\mu_1\beta - \sqrt{2\mu_1\beta\left(||\psi_1||^2_{H^2}\left[1+\frac{p}{H_1}\right] + ||\psi_2||^2_{H^2}\left[1+\frac{p}{H_2}\right]\right)}. \tag{15}$$

As before, we now take the supremum over $U \in A$ with $||U|| = 1$, and then the infimum over all subspaces A of $L^2(\Omega)$ with dimension k. We denote by λ_k the critical points of E and by ν_k the critical points of the functional $G(U) = H_1|\Delta V_1|^2 + H_2|\Delta V_2|^2$ under the constraint K. It follows that for all $\epsilon \in (0,1)$ we get :

$$\lambda_k \geq \mu(1-\epsilon)\nu_k - \frac{1}{2\mu\epsilon}\left(||\psi_1||^2_{H^2}\left[1+\frac{p}{H_1}\right] + ||\psi_2||^2_{H^2}\left[1+\frac{p}{H_2}\right]\right).$$

We now optimize in ϵ. Comparing the resulting estimate with (15), we conclude :

$$\lambda_k \geq max(e_1, e_k)$$

where

$$\begin{aligned} e_1 &= \mu\mu_1\beta - \sqrt{2\mu_1\beta\left(||\psi_1||^2_{H^2}\left[1+\frac{p}{H_1}\right] + ||\psi_2||^2_{H^2}\left[1+\frac{p}{H_2}\right]\right)}, \\ e_k &= \mu\nu_k - \sqrt{2\nu_k\left(||\psi_1||^2_{H^2}\left[1+\frac{p}{H_1}\right] + ||\psi_2||^2_{H^2}\left[1+\frac{p}{H_2}\right]\right)}. \end{aligned}$$

Let us now determine the value of ν_k. We consider the functional $G(U) = H_1|\Delta V_1|^2 + H_2|\Delta V_2|^2$. The critical points of $G(U)$ under the constraint K are the solution of the following eigenvalue problem :

$$\begin{cases} H_1\Delta^2 V_1 = \nu\left(-H_1\Delta V_1 + p(V_1 - V_2)\right) & \text{in} \quad \Omega\times]0,T[, \\ H_2\Delta^2 V_2 = \nu\left(-H_2\Delta V_2 - p(V_1 - V_2)\right) & \text{in} \quad \Omega\times]0,T[, \\ \Delta V_i = 0 & \text{on} \quad \partial\Omega\times]0,T[, \\ V_i = 0 & \text{on} \quad \partial\Omega\times]0,T[, \end{cases} \tag{16}$$

If V is a solution of (16), then $\Phi = H_1V_1 + H_2V_2$ and $\phi = V_1 - V_2$ are respectively solutions of

$$\begin{cases} \Delta^2\Phi = -\nu\Delta\Phi & \text{in} \quad \Omega\times]0,T[, \\ \Delta\Phi = 0 & \text{on} \quad \partial\Omega\times]0,T[, \\ \Phi = 0 & \text{on} \quad \partial\Omega\times]0,T[, \end{cases} \tag{17}$$

and

$$\begin{cases} \Delta^2\phi = -\nu\Delta\phi + \nu\left(\frac{p}{H_1} + \frac{p}{H_2}\right)\phi & \text{in} \quad \Omega\times]0,T[, \\ \Delta\phi = 0 & \text{on} \quad \partial\Omega\times]0,T[, \\ \phi = 0 & \text{on} \quad \partial\Omega\times]0,T[. \end{cases} \tag{18}$$

And then from the first system it follows that $\nu = \mu_k$, while from the second it follows that $\nu = \dfrac{\mu_k^2}{\mu_k + \left(\frac{p}{H_1} + \frac{p}{H_2}\right)}$. So we order $\lambda_1 \leq \lambda_2 \leq, ...,$ and we conclude that

$$\nu_k \in \begin{cases} \mu_i, \ i \in \mathbb{N} \\ \dfrac{\mu_j^2}{\mu_j + \left(\frac{p}{H_1} + \frac{p}{H_2}\right)}, j \in \mathbb{N} \end{cases} \tag{19}$$

We remark that the estimates take into account the inverse of radii of deformation of each layer, namely $r_1 = 0, r_2 = \frac{p}{H_1} + \frac{p}{H_2}$.

7. Conclusion.

The a priori estimates that we have obtained give us not only a bound from below for each eigenvalue, but also a bound from above for the number of negative eigenvalues. These estimates are still valid for a very small coefficient of viscosity.

References

[1] Barkmeijer J. *Local error growth in a barotropic model.* Tellus , 44 A, p314-323, 1992.

[2] Barkmeijer J., Houtekamer P.,An application of finite element methods for modelling wind driven circulations in a stratified ocean Wang X. *Validation of a skill prediction method.* Tellus, to appear.

[3] Bernier C. *Etude et parallélisation d'un code d'éléments finis pour la modélisation quasi-géostrophique des circulations océaniques.*, Thèse INPG, juin 1990.

[4] Bernier C. *Existence of attractor for the quasi-geostrophic approximation of the Navier-Stokes equations and estimate of its dimension.* rapport de recherche INRIA N° 1733, juillet 1992.

[5] Brezis H. *Analyse fonctionnelle.* Masson, 1992.

[6] Chen Y., Chen S., O.Roads J. *Linear predictions and diagnosis of time-averages in a two-level model.* Tellus, 43 A, p 81-96, 1991.
[7] Chow and Holland W.R. *Description of a quasi-geostrophic multi-layer box ocean model.* NCAR internal report, 1986.
[8] Houtekamer P. *Variation of the predictability in a low-order spectral model of the atmospheric circulation.* Tellus, 43 A, p 177-190, 1991.
[9] Dymnikov V.P., Filatov A.N. *Stability of large-scale atmospheric processes.* Leningrad, Gydrometeoizdat, 1990, in russian.
[10] Il'in A. A. *The Navier-Stokes and Euler equations on two-dimensional closed manifolds.* Math. USSR Sbornik, Vol 69 N° 1, 1991.
[11] Kato T. *Perturbation theory for linear operators.* Springer-Verlag, 1966.
[12] Kavian O. *Introduction à la méthode des points critiques.* à paraitre , 1993.
[13] Lacarra J-F., Talagrand O. *Short-range evolution of a small perturbations in a barotropic model.* Tellus, 40 a, p 81-95, 1988.
[14] Le Provost C. *An application of finite element methods for modelling wind driven circulations in a stratified ocean.*, 5th international Conference on finite elements in Water Resources, Springer-Verlag, 1984.
[15] Lions J-L. *Quelques methodes de résolution des problèmes aux limites non linéaires.* Dunod, Paris, 1969.
[16] Temam R. *Infinite-dimensional dynamical systems in mechanics and physics.* Applied Mathematical Sciences 68, Springer-Verlag, 1988.

Application of Fifth-Order Compact Upwind Differencing to Moisture Transport Equation in AGCM

Mikhail A. Tolstykh
Institute of Numerical Mathematics
Russian Academy of Sciences
Leninskij prosp. 32A
117334 Moscow Russia

ABSTRACT. Fifth-order compact upwind differencing possesses very small amplitude and phase errors and may be effectively implemented on MPP due to three-point stencil. Several 1-D and 2-D advection tests were performed including the test on the sphere with the flow over the poles. The used time stepping were implicit third-order Adams scheme and explicit fourth-order low-storage Runge-Kutta method. This scheme was also implemented in atmospheric general circulation model and showed significantly better results than the second-order scheme.

1. Introduction

The development of MPPs renewed the interest to finite-difference techniques in the global atmosphere modelling since these techniques seem to be most suitable for such computers. However, commonly used in this field second- and fourth-order central differencing suffer from significant phase error. Besides, fourth-order central differencing requires five-point stencil.

Let us remind of what is the compact differencing. Routine second-order formula for the computation of the first derivative of the function φ on a mesh with constant step h is written as

$$\left(\frac{\partial\varphi}{\partial x}\right)_i = \frac{\varphi_{i+1} - \varphi_{i-1}}{2h} + O(h^2).$$

At the beginning of the century Numerov suggested another way of computing first derivatives which is called compact differencing

$$\frac{1}{6}\left(\frac{\partial\varphi}{\partial x}\right)_{i-1} + \frac{2}{3}\left(\frac{\partial\varphi}{\partial x}\right)_i + \frac{1}{6}\left(\frac{\partial\varphi}{\partial x}\right)_{i+1} = \frac{\varphi_{i+1} - \varphi_{i-1}}{2h} + O(h^4). \quad (1)$$

The remarkable properties of the formula (1) are that it gives fourth-order accuracy on three-point stencil and besides it is 24 times more accurate than explicit fourth-order central differencing [1]. The phase error of (1) is also significantly better than that of explicit central differencing [1]. However, for waves with lengths $2h$

F.-X. Le Dimet (ed.), *High Performance Computing in the Geosciences*, 159–174.

phase velocity is equal to zero as in many schemes and the formula (1) contains no numerical dissipation to damp these false oscillations.

The certain disadvantage of (1) is also the necessity to solve systems with tridiagonal matrices. However, these matrices are constant and can be inverted in advance. Another possibility is to solve the group of such systems with the help of well parallelized technique, for example cyclic reduction. The choice of these two opportunities obviously depends on the MPP being at hand.

This paper describes application of the recently developed fifth-order compact-upwind differencing (CUD-5) for advection equation in Eulerian formulation associated with the problem of moisture transport in the atmosphere. This problem incorporates very steep, rapidly changing gradients. Besides, the moisture is essentially positive function and erroneous negative values result in false vertical convection in the atmospheric model (AGCM). This problem places rigid constraints on the numerical algorithm. These demands were discussed in [2]. Here we note that in this problem the main goal of the algorithm design is to provide conservation and quasi-monotonicity (i.e. the absence of spurious oscillations) properties simultaneously as well as the absence of negative values. Many works were devoted to this problem (see, for example, [2], [3], [4]).

Although we discuss only Eulerian approach, it is clear that the same technique can be used as some interpolation operator for semi-Lagrangian schemes in spirit of Smolarkiewicz papers [5].

We start with the description of CUD-5 in Section 2, in Section 3 we show the results of 1-D and 2-D benchmarks. The application of CUD-5 to the moisture transport in atmospheric general circulation model (AGCM) is given in Section 4. Finally, Section 5 presents discussion.

2. Fifth-Order Compact Upwind Differencing

The fifth-order compact upwind differencing (CUD-5) was introduced in 1991 [6]. It has remarkable properties of small phase errors and dissipation concentrated on short waves. Its phase error for waves with lengths up to $4h$ (h is grid step) is equal to zero and shorter waves are damped by intrinsic dissipation [7].

This type of spatial approximation is given by [6]

$$\frac{\partial \varphi}{\partial x} = L_h \varphi + O(h^5),$$

$$L_h \varphi = \frac{1}{2h}\left(\Delta(s) - sR^{-1}Q\left(I + \frac{1}{12}\Delta_2\right)^{-1}\Delta_2\right)\varphi, \tag{2}$$

where

$$\Delta_0 f = f_{i+1} - f_{i-1}, \qquad \Delta_2 f = f_{i+1} - 2f_i + f_{i-1},$$

$$\Delta(s) f = \left(\Delta_0 - s\Delta_2\right) f,$$

h is a mesh size.

The operators R and Q are given by

$$R = I + \left(\frac{b}{2} - \frac{s_1}{4}\right)\Delta_0 + \left(\frac{1}{6} - \frac{b}{2}s_1\right)\Delta_2,$$

$$Q = I + \left(\frac{a}{2} - \frac{s_1}{4}\right)\Delta_0 + \left(\frac{1}{6} - \frac{a}{2}s_1\right)\Delta_2,$$

$$a = -\frac{2}{15s}, \qquad b = \frac{1}{5s}, \qquad |s| = \frac{2}{\sqrt{5}}, \qquad |s_1| > 5/24,$$

where I is the identity operator.

We see that CUD-5 consists of the generalized first-order upwind operator $\Delta(s)$ and Pade approximant annihilating corresponding terms in the Taylor expansion series for truncation error. This Pade approximant is based only on tridiagonal operators.

It was proved in [6] that operator L_h is positive for $s > 0$ or negative for $s < 0$ if $\mathbf{sgn}\, s_1 = \mathbf{sgn}\, s$, $|s_1| > 5/24$. The proper choice of s (upwinding) provides the stability of CUD-5 for frozen coefficients approximation.

One can see that this approximation of spatial derivative requires inversion of tridiagonal operators. Dispersion and dissipation properties of these operators are shown on Fig. 1 [8]. Here we note that the parameter s_1 of L_h controls phase error and dissipation for waves in the range of $2h$ to $6h$: the greater s_1, the greater is dissipation, the minimum phase error being for $s_1 = 0.9$. The choice of this parameter has also to a certain extent an effect on stability limit. The cost of this discretization is at least 21 arithmetic operations per grid point.

In the grid points where velocity changes its sign, the direct applying of such spatial discretization leads to the loss of conservation property. To preserve this property, the idea of flux splitting suggested in earlier eighties can be used [9], [10]. The flux splitting leads to increase of the amount of arithmetic operations, but now both three-point operators R and $I + \Delta_2/12$ can be inverted in advance.

3. Advection Tests

Using 1-D advection equation written in the conservative form

$$\frac{\partial q}{\partial t} + \frac{\partial uq}{\partial x} = 0, \tag{3}$$

where u is supposed to be independent from time, we consider two time schemes. They are

- Adams third-order conditionally stable implicit scheme

$$q^{n+1} - q^n = -\tau\left(\frac{5}{12}L_h\left(uq^{n+1}\right) + \frac{8}{12}L_h\left(uq^n\right) - \frac{1}{12}L_h\left(uq^{n-1}\right)\right), \tag{4}$$

- explicit Runge-Kutta fourth-order low-storage method with increased stability [11]

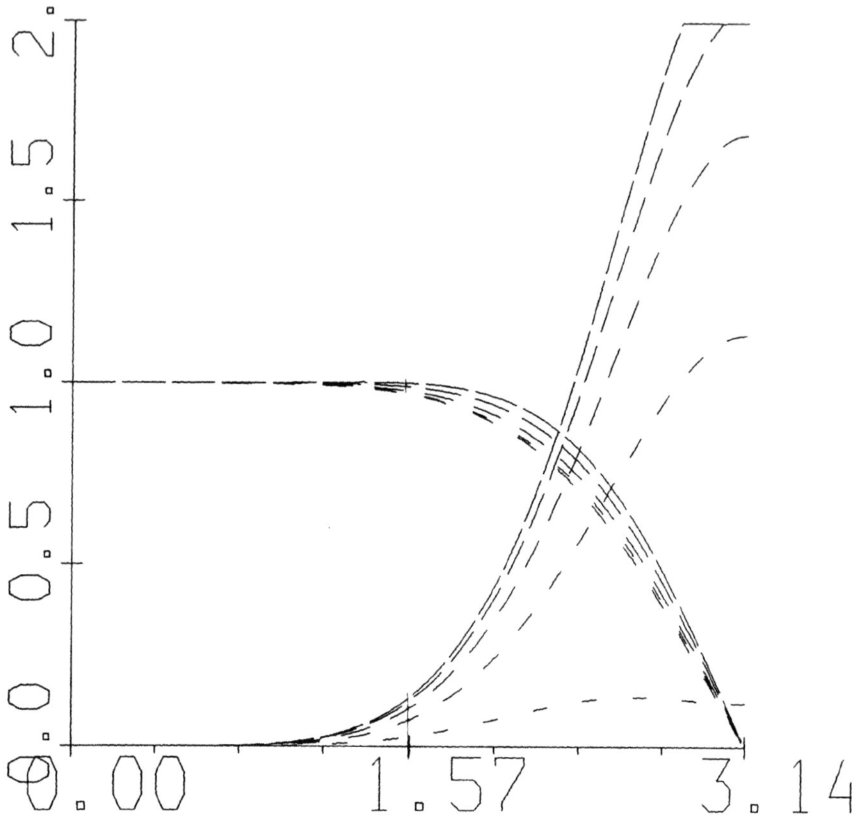

Fig. 1. Dispersion and dissipation properties of CUD-5 for various s_1. Minimum value of s_1 is 0.21, increment is 0.45. The longer dashes correspond to larger values of s_1.

$$\begin{aligned}
q^{(1)} &= L_h\big(uq^n\big),\\
q^{(2)} &= L_h\Big(u\big(q^n + \frac{\tau}{2}q^{(1)}\big)\Big),\\
q^{(3)} &= L_h\Big(u\big(q^n + \frac{\tau}{2}q^{(2)}\big)\Big),\\
q^{(4)} &= L_h\Big(u\big(q^n + (0.5-\alpha)\tau q^{(2)} + \alpha\tau q^{(3)}\big)\Big), \quad \alpha = 24\cdot 4.0869614e-3,\\
q^{(5)} &= L_h\Big(u\big(q^n + \tau q^{(4)}\big)\Big),\\
q^{n+1} &= q^n + \frac{\tau}{6}\Big(q^{(1)} + 2q^{(2)} + 2q^{(3)} + q^{(5)}\Big).
\end{aligned} \tag{5}$$

Obviously, time stepping for use with CUD-5 should be of sufficient accuracy in order not to mask high spatial accuracy. The Adams scheme proved to be the best among other implicit third-order schemes and second-order Krank-Nicholson scheme [12]. However, its implementation on MPP may cause difficulties. The explicit Runge-Kutta schemes seem to be more suitable for MPPs.

Stability estimates for Adams scheme in conjunction with CUD-5 gives the limitation CFL$\leq$ 0.8. However, beyond this limit the amplifying factor is very close to 1[8]. The similar limitation for Runge-Kutta scheme (5) gives maxiimum CFL$\approx$ 1.6, but in this case amplifying factor is significantly greater.

Now we consider 1-D tests suggested in [4] for the spatial discretization (2) in conjunction with time stepping (4) and (5). They are described by Eq. (3) in the periodical interval of 40 points for rectangle and triangle distributions or 80 grid points for Gaussian initial distribution with constant velocity. 100 time steps were performed for rectangle and triangle distributions, while for Gaussian distribution 800 time steps were performed. Implicit time stepping was implemented using GMRES solver with Hausholder transformation [13]. The results of these tests for time stepping (4) and (5) and CFL =0.5 are shown on Fig. 2, A and B respectively. Comparing these figures one can see that Runge-Kutta time stepping provides somewhat better results.

The comparison of different schemes (among them van Leer and Smolarkicwicz schemes) was presented in [4], the best results in this paper being for Piecewise Parabolic Method (PPM). We may note that CUD-5 in the case of Gaussian initial distribution gives better curve than PPM [4] where the maximum decreases to 0.881. For the triangle distribution they are comparable: PPM decreases the maximum to 0.768 without steepening procedure and to 0.828 with steepening, the shape of the distribution in the latter case being similar to rectangle. PPM generates no spurious negative values. In the case of CUD-5 we have the maximum value 0.868 with the single negative value -0.017 for Adams scheme (4) and for Runge-Kutta scheme (5) the maximum value is 0.915 with 3 negative values less than 0.013. Applying Flux Corrected Transport (FCT) [14] periodically one can eliminate the negative value in this test, the maximum value being about 0.75 (not shown). The results of applying CUD-5 for the rectangle distribution are worse than that of PPM since CUD-5 cannot simulate the waves with lengths $2h$ and $3h$ properly. Similarly, periodical applying of FCT eliminates over- and undershoots but smooths the rectangle distribution. The triangle wave and Gaussian distribution are perhaps most appropriate cases in meteorological problems. It is also worth mentioning that the Piecewise Parabolic Method is more expensive than CUD-5.

To investigate the shape-preserving property of CUD-5 operators, 2-D advection test on the sphere with the flow directly over the poles was also carried out. This is the first test from the set of tests for shallow water equations in spherical geometry [15]. This test is important since near the poles meridians of spherical coordinate system converge resulting in local CFL numbers of order 10 near the poles. The test is described by the equation

$$\frac{\partial q}{\partial t} + \frac{1}{a\cos\varphi}\left(\frac{\partial uq}{\partial \lambda} + \frac{\partial v\cos\varphi q}{\partial \varphi}\right) = 0,$$

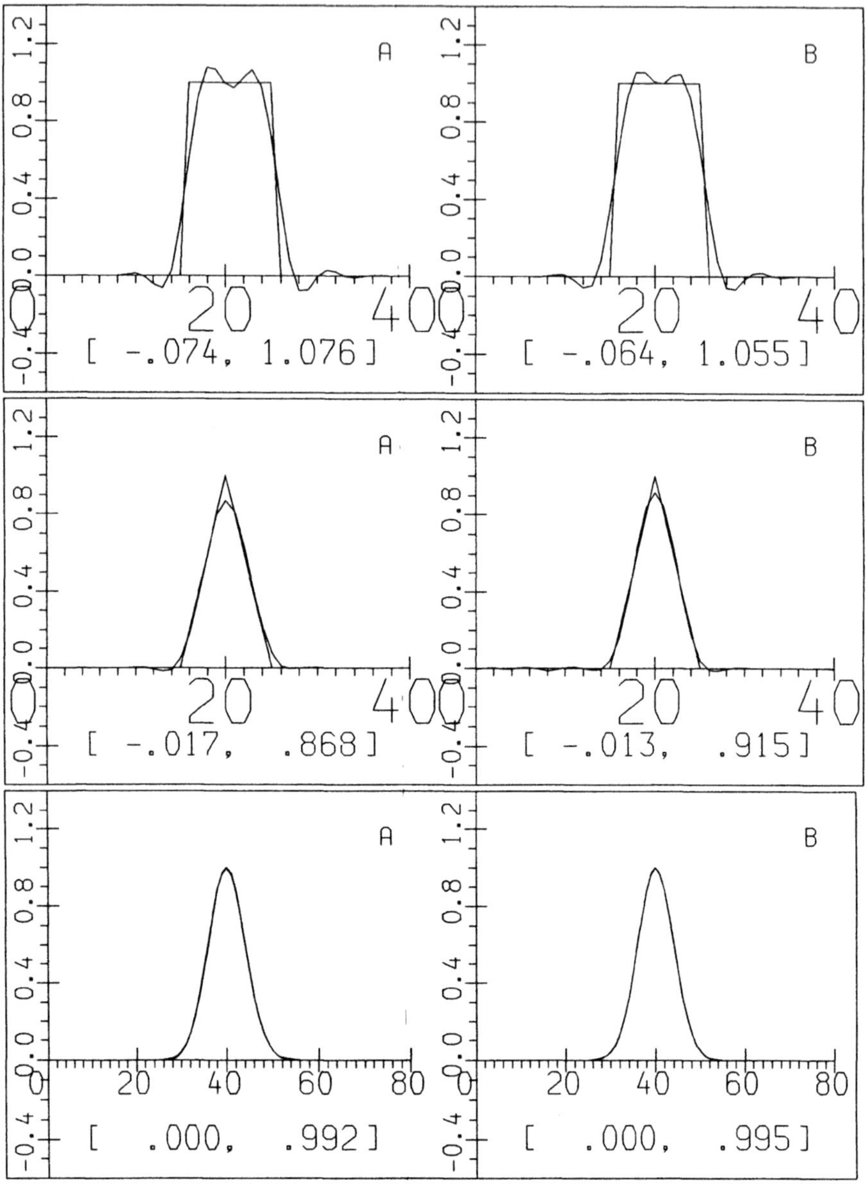

Fig. 2. 1-D advection with CUD-5 in the periodical interval for different types of initial distribution, CFL =0.5: A – implicit time stepping (4), B – explicit Runge-Kutta time stepping (5).

where

$$u = u_0\left(\cos\varphi\ \cos\alpha + \sin\varphi\ \cos\lambda\ \sin\alpha\right),$$

$$v = u_0 \sin\lambda \sin\alpha,$$

$$q(\lambda,\varphi)\Big|_{t=0} = \begin{cases} 500\big(1+\cos(\pi r/R)\big) & \text{if} \quad r < R \\ 0 & \text{if} \quad r \geq R, \end{cases}$$

where a is Earth radius, λ is longitude, φ is latitude, u is λ-component of velocity and v is its φ-component. The radius $R = a/3$ and the wind velocity $u_0 = 2\pi a/(12$ days), r is the great circle distance between $(\lambda,\ \varphi)$ and the center, initially taken as $(\lambda_c,\ \varphi_c) = (3\pi/2,\ 0)$:

$$r = a \arccos\big[\cos\varphi\ \cos(\lambda - 3\pi/2)\big].$$

The initial distribution of q is the cosine bell. It is shown by dashed lines in Fig. 3. The parameter α is the angle between the axis of solid body rotation and the polar axis of the spherical coordinate system. The flow over the poles corresponds to $\alpha = \pi/2$. The results for $\alpha = 0$ corresponding to the flow along equator are of the same quality as for 1-D tests with Gaussian initial distribution and are not shown.

The grid contained 128 points in longitude and 64 points in latitude. These grid parameters are the same as in [5] and some other papers. Near the poles longitudinal Fourier filtering commonly used in atmospheric models was applied (each harmonic was multiplied by the factor depending on latitude and wavenumber).

To ensure conservation property the flux splitting is used. It can be briefly described in the following manner. Suppose we have 1-D advection equation

$$\frac{\partial q}{\partial t} + \frac{\partial uq}{\partial x} = 0$$

and we want to use upwind CUD-5 operators

$$L_h^+ \text{ for } u \geq 0 \qquad \big(s,\ s_1 > 0\big),$$

$$L_h^- \text{ for } u < 0 \qquad \big(s,\ s_1 < 0\big).$$

We may then write

$$\frac{\partial q}{\partial t} + \frac{1}{2}\Big[L_h^+\big((u+C)q\big) + L_h^-\big((u-C)q\big)\Big] = 0,$$

$$C \geq \max|u|.$$

Thus the algorithm becomes conservative [7]. It can be also shown that the operator in brackets is positive for $u = const$ thus providing the stability for the frozen coefficients approximation.

The results are shown in Fig. 3. CFL number near equator was 0.2, 640 time steps were carried out. Now we can compare these results with other schemes in Eulerian formulation [5]. In the paper [5] this number was 0.05 for MPDATA scheme. This means that "effective" CFL number in the case of CUD-5 in conjunction with implicit time stepping (4) is somewhat less than for MPDATA explicit algorithm. Comparing the shape of the solution after one revolution, we may see that it is better than the shape of distribution of MPDATA solution both in monotone and non-monotone versions.

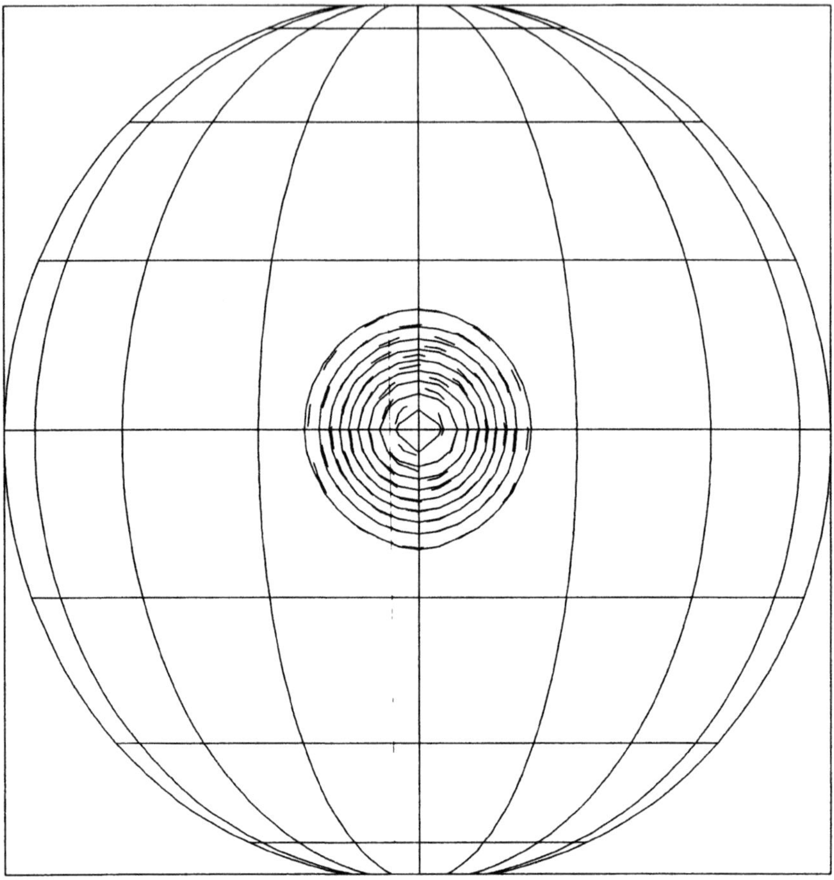

Fig. 3. 2-D advection on the sphere with the flow over the poles for CUD-5 and time stepping (4), CFL=0.2. The initial distribution is plotted by dashed lines, contour interval is 100 with minimum contour value 0.

Now we compare normalized errors for maximum value (E_{max}), minimum value (E_{min}) and global mean value (E_0) [15] with those for MPDATA scheme. These errors at each time step are defined as follows

$$E_{max} = \frac{\max_{\lambda,\varphi} q - \max_{\lambda,\varphi} q_0}{\max_{\lambda,\varphi} q_0},$$

$$E_{min} = \frac{\min_{\lambda,\varphi} q}{\max_{\lambda,\varphi} q_0},$$

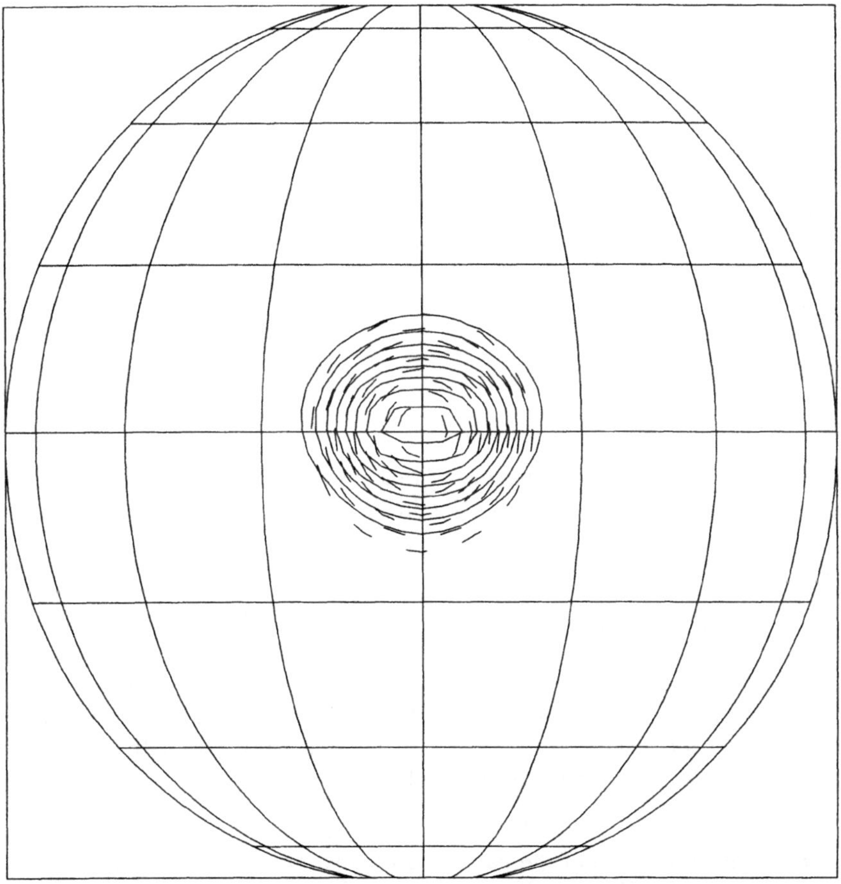

Fig. 4. 2-D advection on the sphere with the flow over the poles for CUD-5 and time stepping (4), CFL=0.33. Plotting conventions are the same as in Fig. 3.

$$E_0 = \frac{\sum\limits_{\lambda,\varphi} q \cos\varphi - \sum\limits_{\lambda,\varphi} q_0 \cos\varphi}{\sum\limits_{\lambda,\varphi} q_0 \cos\varphi},$$

where q_0 is the initial distribution and q is the distribution at the given time step.

Here we present time-averaged and maximum values of these errors. They are given in Table 1.

Comparing these errors with both monotone and non-monotone version of MPDATA scheme, we may conclude that the error for minimum value (2 per cent) is absent in MPDATA but the error for maximum value in the case of CUD-5 is one order of magnitude less. The error for global mean value (less than 1 per cent) is

TABLE I
Errors for 2-D test with different time stepping and CFL numbers.

Time stepping	CFL	$\bar{E}_{max}$	$\bar{E}_{min}$	$\bar{E}_0$	max E_{max}	max E_{min}	max E_0
Implicit (4)	0.2	-0.033	-0.019	-0.0097	-0.052	-0.029	-0.018
Implicit (4)	0.33	-0.075	-0.035	-0.043	-0.152	-0.061	-0.082
Explicit (5)	0.08	-0.031	-0.011	-0.0003	-0.060	-0.018	0.0014

also absent in MPDATA. The presence of this error for CUD-5 is explained by longitudinal filtering near the poles. On the first quarter of the trajectory (i.e. before the pole) this error is absent. Using more exact Runge-Kutta technique decreases the error by the order of magnitude.

It is also worth investigating the maximum achievable CFL number near equator for implicit time stepping (4). It was found to be equal to 0.33. The resulting pattern is shown in Fig. 4. We may conclude that in this case CUD-5 gives approximately the same quality of distribution as in MPDATA algorithm, but now the error for maximum value is the same as for the non-monotone version of MPDATA and one half of that for the monotone version of MPDATA.

Considering this test for CUD-5 in conjunction with Runge-Kutta scheme (5), we may see that the shape of the pattern is approximately the same as in the case of implicit scheme (4) (Fig. 5). The maximum achievable CFL number in this case is 0.08. The error for maximum and minimum values are somewhat better than for implicit time stepping (4), the error for global mean value is significantly less than for scheme (4).

Comparing results for CUD-5 with different semi-Lagrangian schemes given in [5] and [16], we can make several observations. The first is that the shape of the distribution in the case of CUD-5 is not worse than that for the best of shape-preserving semi-Lagrangian tests. The second is that error measures of CUD-5 except for global mean value are worse in comparison with these schemes [16]. However, this test masks two sources of errors intrinsic for semi-Lagrangian schemes. The first is the accuracy of estimation of departure points and can be made negligible. But the most important source of errors is the absence of conservation property which result in "noisy" solutions in the case of non-uniform flow. This case makes no problem for any conservative Eulerian algorithm of sufficient accuracy. Besides, the dominant direction of atmospheric flow is not along the meridian but along the equator. Thus the fact that semi-Lagrangian schemes provide better results for this test does not mean that these algorithms are better in general case.

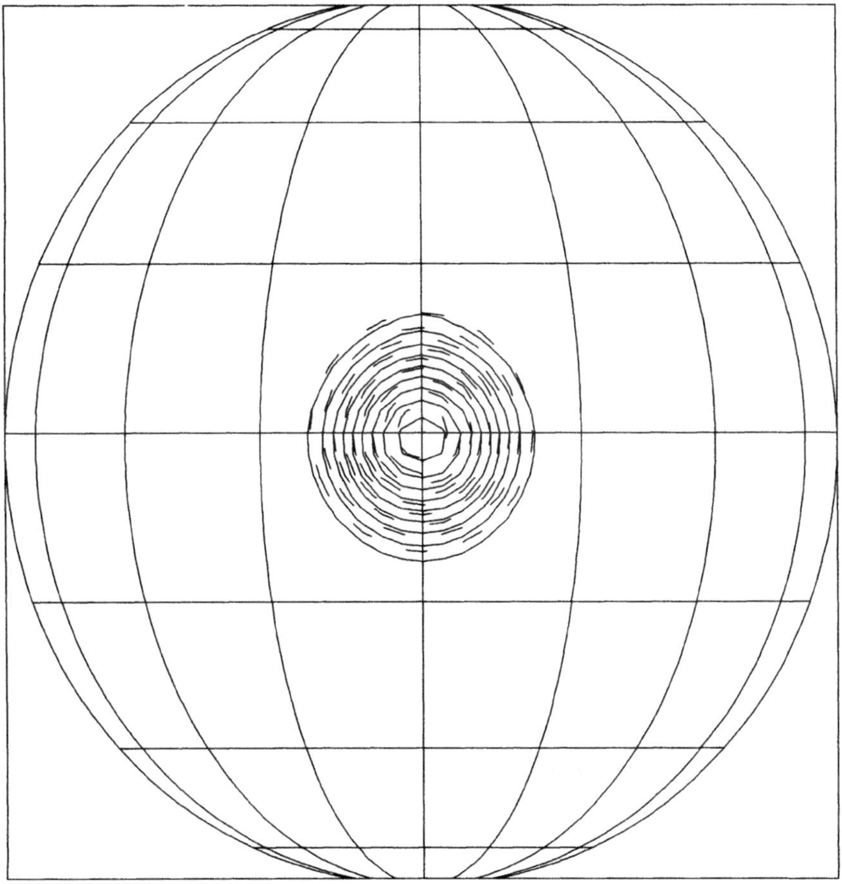

Fig. 5. 2-D advection on the sphere with the flow over the poles for CUD-5 and Runge-Kutta time stepping (5), CFL=0.08. Plotting conventions are the same as in Fig. 3.

4. The Use of CUD-5 for Moisture Transport Equation in AGCM

In the previous ten years the spectral method played the dominant role in numerical modelling of the global atmosphere. Despite its conservation property, this method suffers from certain disadvantages:

- it generates spurious oscillations due to Gibbs phenomena;
- it is computationally ineffective, since while increasing spatial resolution the amount of arithmetic operations increases as $O(N_1 \log_2 N_1 N_2^2 N_3)$, where N_1 is the number of points in longitudinal direction, N_2 – in latitude and N_3 is the number of vertical levels in AGCM.

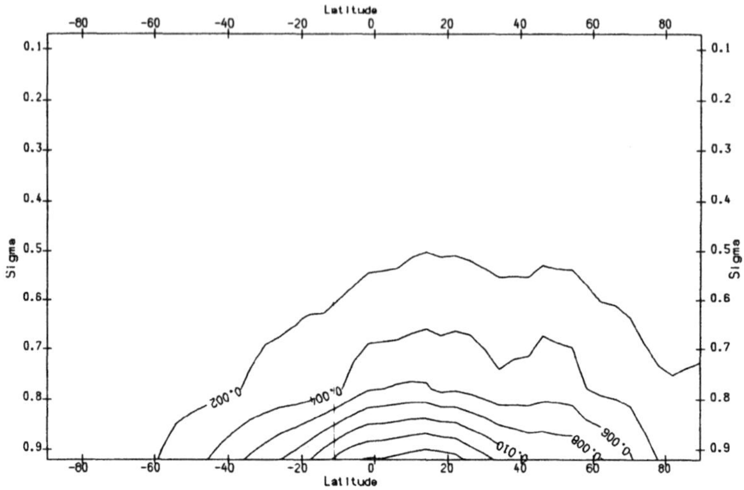

Fig. 6. The time-average zonal mean humidity for the first experiment (second-order horizontal discretization, explicit time scheme).

These facts stimulated the development of new generation techniques for moisture transport. The semi-Lagrangian moisture transport after the excellent works of Rasch and Williamson [2] is now the most popular method. Semi-Lagrangian technique has no intrinsic conservation property, so this method enforces conservation property via special fixer applied at each time step.

The idea was to create a competitive conservative Eulerian finite-difference in space and time solver for moisture transport in AGCM using CUD-5 and implicit third-order time stepping.

As a basic model, the AGCM of the Institute of Numerical Mathematics (INM), Russian Academy of Sciences, was used. The numerical method used in this model is second order on the staggered grid "C" Arakawa-like scheme in space with explicit Matsuno time stepping, $\Delta t = 6\ min$. The spatial resolution is 5^0 in longitude, 4^0 in latitude, 7 equally spaced σ-levels.

This model produces rather realistic climatology, as its participation in the Atmospheric Models Intercomparison Project (AMIP) has shown. But as far as moisture concerned, there were some discrepancies in zonal mean fields (i.e. the fields averaged over longitude) as well as in seasonally averaged distributions.

Now we describe the new version of moisture transport algorithm in AGCM of INM. The moisture transport equation in AGCM is given as

$$\frac{\partial p_s q}{\partial t} + \frac{1}{a \cos\varphi}\left(\frac{\partial p_s u q}{\partial \lambda} + \frac{\partial p_s v \cos\varphi q}{\partial \varphi}\right) + \frac{\partial p_s \dot{\sigma} q}{\partial \sigma} = F, \tag{6}$$

where $\sigma = p/p_s$ is the vertical coordinate (p_s is sea-level pressure), F describes sinks and sources of moisture due to condensation and evaporation.

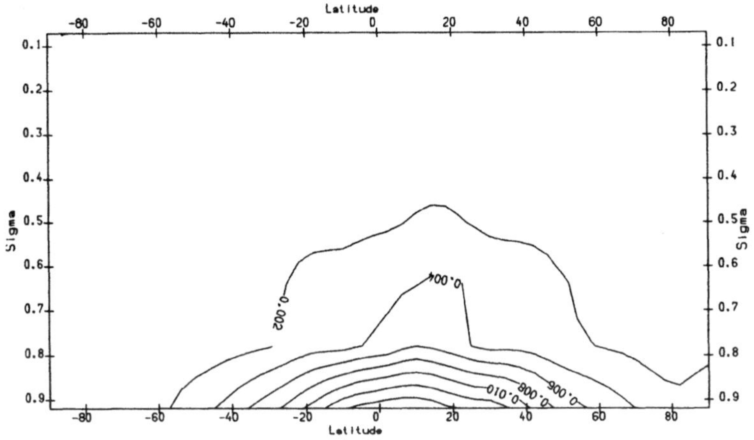

Fig. 7. The time-average zonal mean humidity for the second experiment (CUD-5 horizontal discretization, implicit time scheme(4)).

For approximation of space derivatives in horizontal directions CUD-5 discretization (2) is used. Vertical discretization is routine second-order central differencing on the staggered grid.

The implicit time stepping via third-order conditionally stable scheme (4) was implemented with the help of GMRES solver with Hausholder transformations [13]. Simple scaling of the matrix was used for faster convergence of iterations. It takes no more than 8 iterations (2 of them with the simple operator $\Delta(s)$) for convergence with the relative accuracy 10^{-4}.

Two experiments were carried out. The first one was with the old version of moisture transport (i.e. with second order central differencing and explicit time stepping), $\Delta t = 6\ min$. The second experiment was with CUD-5 and the third-order implicit time stepping (4), $\Delta t = 60\ min$. Initial conditions for AGCM were 1.06.1987 real data of the European Center for Medium-Range Weather Forecasts (ECMWF). The AGCM was integrated for 95 days. Here we describe only the results for moisture. Time-averaged zonal mean (i. e. averaged over longitude) humidity field for the first (control) experiment is presented on Fig. 6, while Fig. 7 shows this field for the fifth-order experiment. We may see that for the second experiment the humidity field is much closer to the observed field. The false extremum in mid-latitudes vanished, the whole picture is more smooth. However, upon inspecting Fig. 7 we see rather steep gradient near 20^0. This gradient is explained by rather coarse vertical mesh (7 equally spaced σ-levels) and central differencing by σ-coordinate. The analysis of vertical profiles of moisture confirms this observation. Thus an improvement in the vertical transport of moisture is also needed for the current version of the model. Comparing the geographical time-averaged distribu-

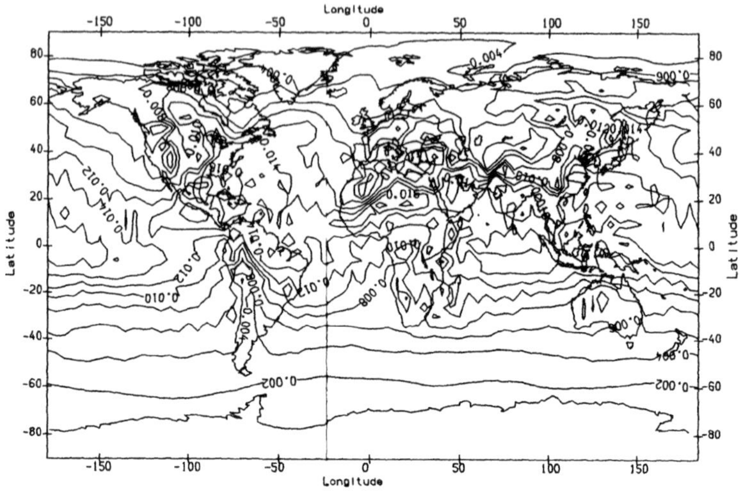

Fig. 8. The time-average humidity field at $\sigma = 0.92$ for the first experiment (second-order spatial discretization, explicit time scheme).

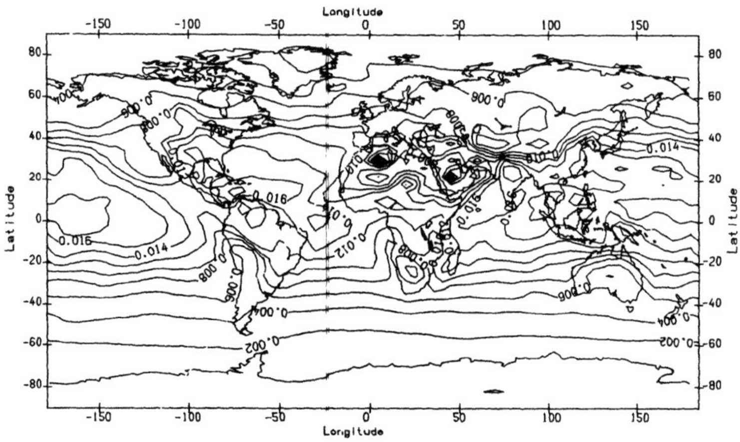

Fig. 9. The time-average humidity field at $\sigma = 0.92$ for the second experiment (CUD-5, implicit time scheme (4)).

tion of humidity at the level $\sigma = 0.92$ for the control and fifth-order experiments (Figs. 8 and 9 respectively), we may see that the whole picture looks much better in the case of CUD-5. Hence, we may conclude that fifth-order compact upwind differencing is appropriate when solving moisture transport equation.

5. Discussion

The fifth-order compact upwind differencing used in this paper possesses the following favorable properties:

1. small phase and amplitude errors,
2. conservation property,
3. compact (three-point) stencil of approximation.

These properties were verified with several 1-D and 2-D tests for Eulerian advection equation. The implementation of this method for moisture transport in the general circulation model of the atmosphere has shown that it works much better than previously used second-order explicit scheme. However, this method is rather costly. To increase its computational efficiency for unsteady problems time stepping should allow computations with CFL number about 1 without significant errors due to time stepping. Third-order implicit conditionally stable scheme used in this study has shown quite satisfactory results. However, it may be difficult to implement it on MPPs. Explicit Runge-Kutta scheme tested in this paper gives good accuracy and seems to be more suitable for MPPs. Another approach is to use CUD-5 as some interpolation operator in semi-Lagrangian schemes.

Acknowledgements

The author thanks his father, Prof. A.I.Tolstykh, for helpful discussions. The idea of this work was strongly stimulated by his researches. I am also very grateful to Prof. F.-X. LeDimet for support of my participation in the Workshop "High Performance Computing in Geosciences".

References

[1] M. J. P. Cullen, Finite-elements method, in *Numerical methods used in atmosherical models*, GARP Publications Series No.17, Vol. 2, 1979.

[2] P. J. Rasch, D. L. Williamson, Computational aspects of moisture transport in global model of the atmosphere, *Quart. J. Roy. Meteorol. Soc.* **116**, 1071 (1990).

[3] P. Smolarkiewicz, W. W. Grabowski, The multidimensional positive-definite advection transport algorithm: Nonoscillatory option, *J. Comput. Phys.* **86**, 355 (1990).

[4] R. L. Carpenter, K. K. Droegemeier, P. R. Woodward, C. E. Hane, Application of the Piecewise Parabolic Method (PPM) to meteorological modelling, *Mon. Weather Rev.* **118**, 586 (1990).

[5] P. K. Smolarkiewicz, P. J. Rasch, Monotone advection on the sphere: an Eulerian versus semi-Lagrangian approach, *J. Atmos. Sci.* **48** 793 (1991).

[6] A. I. Tolstykh, On a class of noncentered compact difference schemes of fifth-order based on Pade approximants, *Sov. Math. Dokl.* **44**, 69 (1992).

[7] A. I. Tolstykh, *High accuracy non-centered compact difference schemes for fluid dynamics applications* (to be published, World Scientific, New York, 1993)

[8] M. A. Tolstykh, Application of fifth-order compact upwind differencing to moisture transport equation in atmosphere, *J. Comput. Phys.*, submitted for publication.

[9] J. L. Steger, R. F. Warming, Flux-vector splitting of the inviscid gasdynamics equations with application to finite difference methods, *J. Comput. Phys.* **40**, 263 (1981).

[10] B. van Leer, Flux-vector splitting for the Euler equations, *Lecture Notes Phys.* **170**, 507 (1982).

[11] P. J. Van der Houwen, Explicit Runge-Kutta formulas with increased stability boundaries, *Numer. Math.* **20** (1972), 149-164.

[12] M. A. Tolstykh, High-order finite-difference techniques in space and time for moisture transport equation in AGCM, in *Numerical Processes and Systems*, **10**, (Nauka, Moscow, 1993) (to be published, in Russian).

[13] H. F. Walker, Implementation of the GMRES method using Hausholder transformations, *SIAM J. Sci. Stat. Comput.* **9**, 152 (1988).

[14] S. T. Zalesak, Fully multidimensional flux-corrected transport algorithms for fluids, *J. Comput. Phys.* **31**, 335 (1979).

[15] D. L. Williamson, J. B. Drake, J. J. Hack, R. Jakob, P. N. Swarztrauber, A standard test set for numerical approximations to the shallow water equations in spherical geometry, *J. Comput. Phys.* **102** 211 (1992).

[16] D. L. Williamson, P. J. Rasch, Two-dimensional semi-Lagrangian transport with shape-preserving interpolation, *Mon. Weather Rev.* **117** 102 (1989).

A PARALLEL ALGORITHM FOR SIMULATION OF LONG RANGE TRANSPORT OF AIR POLLUTION

Roar Skålin
SINTEF Industrial Mathematics
N-7034 Trondheim
Norway
Ivar Lie
NDRE
Div. for Electronics
P.O.box 25
N-2007 Kjeller
Norway

and

Erik Berge
EMEP/MSC-W
The Norwegian Meteorological Institute
Postbox 43
Blindern
N-0313 Oslo
Norway

ABSTRACT. The simulation of long range air pollution using regional scale models requires a substantial amount of CPU-time on traditional supercomputers and should be able to benefit substantially from effective parallel algorithms. In this paper we describe a continuous domain decomposition algorithm for advection-diffusion equations. The algorithm is applied to an air pollution model, involving advection in the horizontal directions and advection-diffusion and chemical reactions in the vertical direction. The parallel model has been implemented on the Intel Paragon. The numerical experiments shows that the parallel algorithm scales well and the results for the full parallel model are encouraging with respect to the possibilities for operational use.

1. Introduction

The simulation of long range air pollution involves the solution of an advection-diffusion problem in three space dimensions combined with chemical reactions. Using regional scale models, these problems are often simulated for periods up to several years. The resulting computational task may require months of CPU time on traditional supercomputers, and these simulations should therefore be able to benefit substantially from effective parallel algorithms.

F.-X. Le Dimet (ed.), High Performance Computing in the Geosciences, 175–185.

We have earlier studied a continuous multidomain algorithm for advection [6], [9] and advection-diffusion [7], [8] equations, and discussed how these algorithms can be applied to atmospheric models. In this paper we report an implementation of the algorithm in an Eulerian air pollution model. We have chosen to parallelise an existing model currently being developed at the EMEP/MSC-W (EMEP = Co-operative programme for monitoring and evaluation of the long range transmission of air pollution in Europe, MSC-W = Meteorological Synthesizing Centre West), which is hosted at the Norwegian Meteorological Institute. The model will be described in detail in a forthcoming report. The aim of the model is to calculate operationally monthly and annual transport of air pollution in Europe. The model has a high horizontal resolution (50km*50km) covering a large geographical domain extending from the eastern part of North-America to east of the Ural mountains in Russia, and from Northern Africa in the south to arctic regions beyond the North Pole. Such a model will support already existing model tools employed during many years at EMEP/MSC-W [14]. It is important to notice that only a preliminary version of the Eulerian model is going to be applied in this study. However, we consider the physical and chemical parts of this model enough developed to be a realistic basis for experiments with parallel computing. Using the domain decomposition idea developed in [6] and [8] together with the Eulerian air pollution model, this approach gives a useful result, namely an operational parallel model.

The domain decomposition algorithm developed in [6] and [8] is for the continuous problem, and hence not tied to a specific discretisation scheme. However, when applying a specific discretisation scheme, we have to show that the problem is well-posed and that the solution converges to the solution of the continuous problem. This is done for spectral collocation in [6]. The discretisation scheme used in the DNMI model is a flux-based scheme derived by Bott [1], [2]. For such schemes we have two alternatives using a domain decomposition algorithm, either based on the physical variables as described in section 2 below, or by using a flux-balance technique, see e.g. [10]. At this point it is difficult to say which of these alternatives is the most efficient. A comparison of different discretisation schemes and domain decomposition techniques will appear in a forthcoming paper.

The outline of the paper is as follows. In Section 2 we review the continuous domain decomposition algorithm with special emphasis on versions suitable for atmospheric model. We then describe the Eulerian air pollution model and the parallelisation in Section 3 and 4 respectively. Most of the parallel experiments have been carried out on an Intel Paragon, and in section 4 we also review the Paragon architecture and discuss some implementation issues on this system. Section 5 contains results from the numerical experiments.

2. The Continuous Multidomain Algorithm

2.1. THE ADVECTION EQUATION

We first consider the advection equation

$$\psi_t + \nabla \cdot (\mathbf{u}\psi) = 0 \tag{1}$$

where ψ is the scalar to be advected and $\mathbf{u}$ is the velocity field.

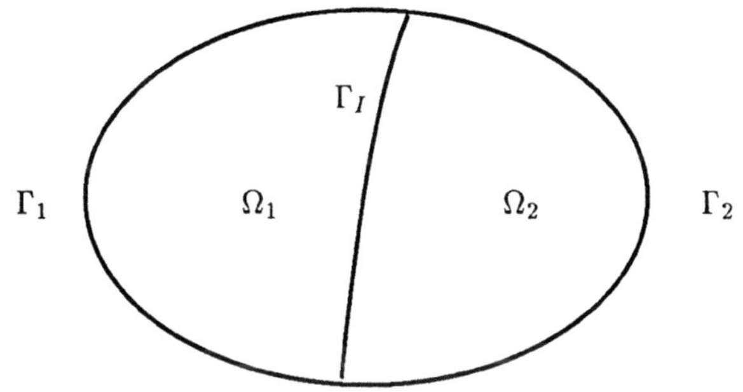

Fig. 1. The two-domain case

Now consider a 2D domain Ω and divide it into two subdomains Ω_1 and Ω_2, see Figure 1. The interface is denoted by Γ and let $\mathbf{n}_1$ and $\mathbf{n}_2$ be normal vectors on $\partial\Omega_1 \setminus \Gamma$ and $\partial\Omega_2 \setminus \Gamma$ respectively. At the interface we construct a normal vector $\mathbf{n}_\Gamma$ pointing from Ω_1 into Ω_2. Given a velocity field, we compute normal velocities:

$$u_n|_{\Omega_1} = \mathbf{u} \cdot \mathbf{n}_1 \qquad \text{on} \quad \partial\Omega_1 \setminus \Gamma$$
$$u_n|_{\Omega_2} = \mathbf{u} \cdot \mathbf{n}_2 \qquad \text{on} \quad \partial\Omega_2 \setminus \Gamma$$

The boundary treatment for hyperbolic systems is often done as correction procedures, i.e. one computes all values in a domain from its interior values and then corrects on the boundary after each time step. The correction procedure is based on the local flow direction:

DEFINITION 1. *The open boundary conditions are computed as follows:*
- *If $u_n|_{\Omega_i} > 0$ (outflow) then compute values from within Ω_i.*
- *If $u_n|_{\Omega_i} < 0$ (inflow) then compute boundary values from outside Ω_i.*

DEFINITION 2. *The interface conditions between Ω_1 and Ω_2 with the normal vector $\mathbf{n}_\Gamma$ as shown in Figure 1 are computed as follows:*
- *If $u_n|_\Gamma > 0$ use values from Ω_1.*
- *If $u_n|_\Gamma < 0$ use values computed from Ω_2.*

where $u_n|_\Gamma = \mathbf{u} \cdot \mathbf{n}_\Gamma$.

The use of these boundary and interface conditions requires that Ω and Ω_i are sufficiently smooth such that we can define the normal vector uniquely. The obvious question is then what to do at corner points, and a common technique is to use the velocity component along the angle bisector and decide upon inflow or outflow based on that. In practice however, the exact method for the corner points seems to be of minor importance for the solution of the advection equation and one may just extend one of the edges of the domain to the corner points.

Extension of this algorithm to 3D and multiple domains is straightforward. It is shown in [6] that the resulting problem is well-posed.

2.2. THE ADVECTION-DIFFUSION EQUATION

We discuss a special case of the advection-diffusion equation with diffusion only in the vertical direction. As we will see in the next section, this is the governing equation for the air pollution model. The equation is given by

$$\psi_t + \nabla \cdot (\mathbf{u}\psi) = \frac{\partial}{\partial z}\left(\nu(x,y,z)\frac{\partial \psi}{\partial z}\right) + f(\psi, x, y, z, t) \quad (2)$$

where $\nu(x,y,z)$ is the diffusion parameter, $f(\psi,x,y,z,t)$ a source term and the other symbols have the same interpretation as above. As we will see in the next section, this is the governing equation for the air pollution model.

In this case we consider a 3D domain divided into two subdomains by a horizontal interface, as shown in figure 2.

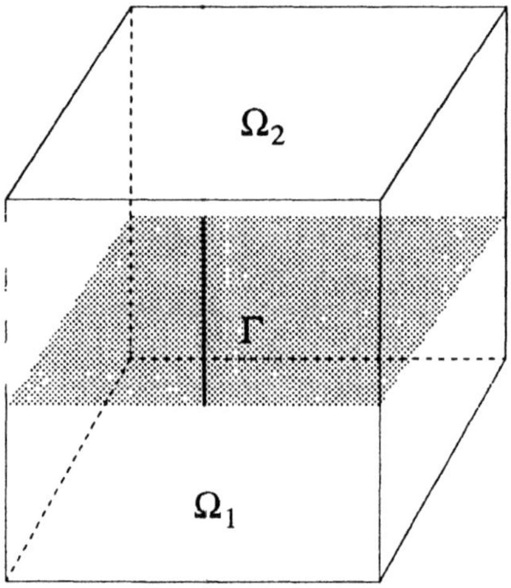

Fig. 2. The domain in the 3D case

The vertical boundaries are considered as open, and since the equation is hyperbolic in the $x-$ and $y-$direction, we can use the boundary conditions given in section 2.1. We also use the hyperbolic open boundary conditions on the edges between the vertical boundaries, and from [6] we can infer that the boundary conditions on the edges can be the conditions imposed on one of the adjacent faces.

Let $\bar{\psi}$ refer to ψ in Eq.2 with zero initial conditions, and let

$$\mathbf{u} \cdot \nabla = c_1(x,y,z)\partial_x + c_2(x,y,z)\partial_y + c_3(x,y,z)\partial_z \quad (3)$$

so for boundaries and interfaces in the positive x-, y-, and z-direction $c_i > 0$ means outflow and $c_i < 0$ means inflow (the term $\nabla \mathbf{u} \cdot \psi$ does not effect the boundary and interface conditions). Now consider the upper boundary of Ω. In this direction the equation is parabolic, and it is shown in [8] that the open boundary conditions are:

$$\nu \frac{\partial \bar{\psi}_2}{\partial z} = 0 \quad \text{when } c_3 > 0 \tag{4a}$$

$$\nu \frac{\partial \bar{\psi}_2}{\partial z} = c_3 \bar{\psi}_2 \quad \text{when } c_3 < 0 \tag{4b}$$

On the lower boundary of Ω we will have $c_3 = 0$, and hence in terms of advection no inflow or outflow occurs. However, we want to impose a boundary condition that allows for deposition, wet or dry, on that boundary. If we use an open boundary condition ("transparent" for deposition) we will have

$$\frac{\partial \bar{\psi}_1}{\partial z} = 0$$

and this can be thought of as an outflow condition. The deposition can be introduced in at least two ways:

- Gravity as an effect in the diffusive flux.
  This boils down to having a term $\alpha \frac{\partial \psi_1}{\partial z} g$, $\alpha \in \mathbb{R}$ in the equation. This term modifies the advection field, so that $c_3 - \alpha g < 0$ at the lower boundary (outflow).
- Sink term in the equation.
  This term will have the form $\alpha \psi_1 u(t) \mathcal{V}(x, y, z)$, $\alpha \in \mathbb{R}$, where $u(t)$ is a time factor, and $\mathcal{V}(x, y, z)$ is a space factor. The space factor will e.g. depend on the state of mixing in Ω, and the time factor will depend e.g. on the time of the day. Both these factors will necessarily be empirical.

Finally, we consider the interface Γ. According to [5, section 1.2] and [7] the interface conditions are given by

$$\nu \frac{\partial \bar{\psi}_1}{\partial z} = \nu \frac{\partial \bar{\psi}_2}{\partial z} \tag{5a}$$

$$\bar{\psi}_1 = \bar{\psi}_2 \tag{5b}$$

Corners and edges must be treated in detail both for interfaces and boundaries. We refer to [8] for a discussion of this issue.

It is straightforward to subdivide the 3D domain also vertically, using the interface conditions given in section 2.1 for the advection equation.

3. The Eulerian Air Pollution Model

3.1. MODEL FORMULATION

The Eulerian air pollution model is based on the following equation for transport, sources (Q_s) and sinks (Q_l) for an air pollutant ψ

$$\psi_t = -\nabla \cdot (\mathbf{u}\psi) + \frac{\partial}{\partial z}\left(k_z \frac{\partial \psi}{\partial z}\right) + Q_s - Q_l \tag{6}$$

As in the previous section, the first term on the right hand side represents advection, while the second term is the vertical diffusion where the vertical diffusivity now is denoted k_z. In our case ψ represents either sulphurdioxide (SO_2) or sulphate (SO_4). Q_s consists of emission for SO_2 and SO_4 and chemical transformation for SO_4, while the chemical transformation is a sink for SO_2. In addition dry and wet deposition are sinks for both SO_2 and SO_4. The wet deposition is the integrated wet scavenging in a column. The dry deposition is applied to the lower boundary where the deposition flux is estimated by use of vertical turbulence profiles and deposition velocities. At the lateral boundaries sulphur estimates from a hemispheric model [15] are used. At the upper boundary (100 hPa) both advective and diffusive fluxes are set to zero. The basic meteorological data (temperature, wind, pressure and humidity) are taken from the Norwegian Weather Prediction Model [4], [12]. Data assimilation is done four times per day, employing interpolation to obtain finer time resolution. The vertical diffusivity k_z is derived from the basic meteorological data.

3.2. DISCRETISATION

We use the time splitting method given in [11] for time-discretisation of Eq. 6. In each coordinate direction the Bott scheme [1], [2] is used for space discretisation of the advection. This scheme conserves mass and is positive definite. The Bott scheme computes the fluxes in and out of grid-boxes by using a local Lagrange interpolant of the concentration field. A family of schemes can be defined by varying the number of points in the interpolant. Five values of ψ are used in the model. When the flow direction is in the positive coordinate direction, the points are symmetrically distributed around the actual grid point. When the flow is in the negative coordinate direction, the interpolation points are shifted one point upstream. The advection split step is explicit in time, whereas the vertical diffusion is implicit.

4. Parallelisation

To parallelise the model, we have so far used a two-dimensional latitude/longitude domain decomposition, with subdomains consisting of contiguous vertical columns extending from the Earth's surface to the tropopause (100 hPa), as shown in figure 3. This approach correspond to using the multidomain algorithm for advection equations given in section 2.1, which is clearly easier than using the algorithm for advection-diffusion equations. It is therefore a sound starting point for the parallelisation effort, but we intend to implement the multidomain advection-diffusion algorithm within the framework of the model as well.

Furthermore, we have chosen to use the Bott scheme for spatial discretisation, as in the sequential version of the model. From the description in section 3.2 we observe that this discretisation requires the values of the concentration field in up to three (possibly artificial) grid points in each direction outside the boundary of each subdomain. Rather than exchanging normal velocities between neighbouring subdomains and extrapolate to find these values, we have chosen to exchange the values of the concentration field. This is illustrated in figure 4 for a horizontal cross-section of the domain.

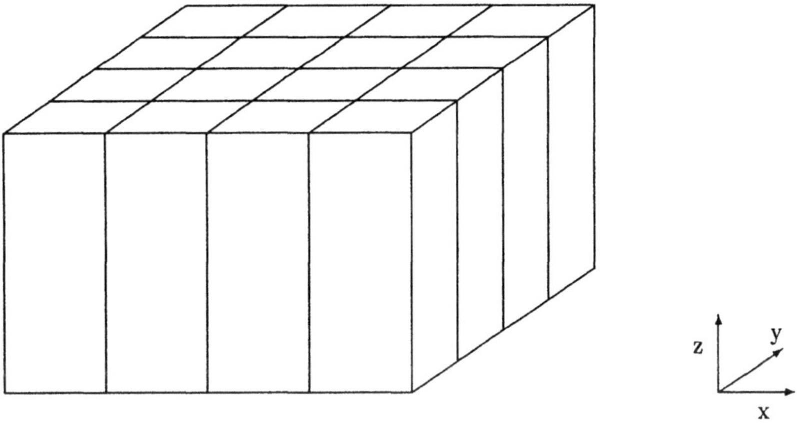

Fig. 3. The physical domain splitted into columns

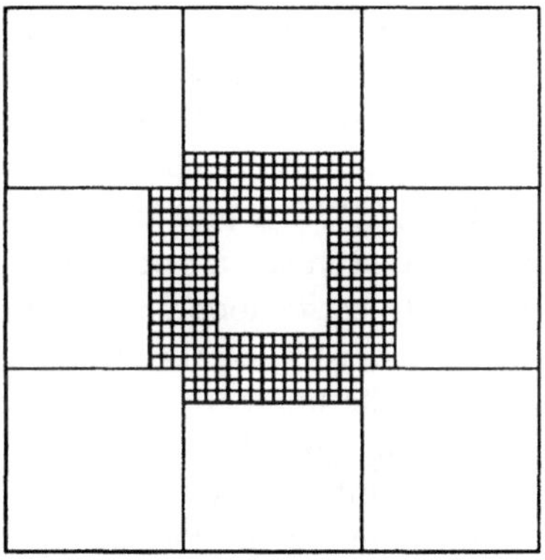

Fig. 4. Area of neighbour-to-neighbour communication for a subdomain

The model has been implemented on the Intel Paragon. The Paragon is a distributed memory MIMD machine with processing nodes arranged in a two-dimensional grid. The data network is constructed from Paragon Mesh Routing Chips (iMRCs) which are connected by high-speed channels. To each iMRC, one node may be attached. The interconnection network is shown in figure 5. The peak bandwidth between two nodes is 200 MB/s in both directions and the theoretical start-up latency for a message issued by a program written in high-level language is 30μs. With the current release of the operating system (Paragon OSF/1 release 1.0) the obtainable values are in the area of 20 MB/s for the bandwidth and 130

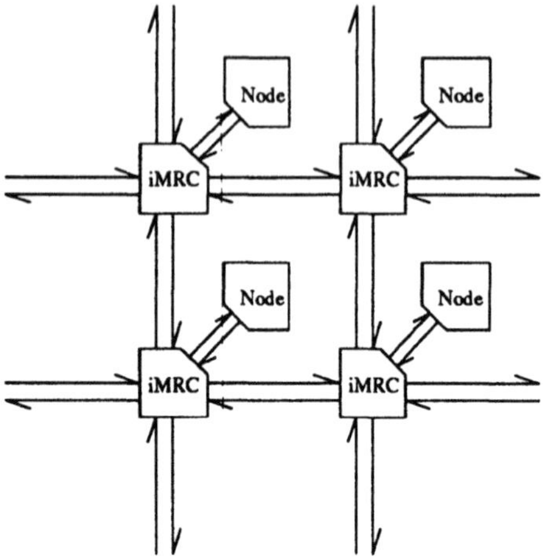

Fig. 5. The Paragon interconnection network

μs for the latency. The processing nodes consists of an Intel i860 XP microprocessor and (currently) 16 MB RAM. The peak performance of the i860 XP is 75 MFLOPS (64-bit arithmetic). Experience with previous generation i860 processors has shown that it is hard to obtain values close to peak on this architecture [3], and performance in the area of 10 MFLOPS is probably more realistic.

The parallelisation strategy used in the air pollution model maps directly onto the Paragon grid architecture with each subdomain assigned to one node. To avoid other communication than neighbour-to-neighbour, we have limited the number of subdomains in each direction to the number of processing nodes available in the corresponding direction on the Paragon. This should ensure that communication time does not increase when the number of subdomains are increased. We have used the native communication primitives on the Paragon, and asynchronous message-passing is used to overlap communication and computation as much as possible.

The air pollution model is relatively I/O intensive and I/O is therefore an important issue for a successful parallel implementation of the operational model. The I/O is currently done through one of the compute nodes. This node reads the meteorological fields from a file and distributes the data to the other nodes. It also collects the results and writes to a file. Use of the Parallel File System (PFS) [13] on the Paragon will allow for a more efficient I/O-handling. This has not yet been implemented in the model, partially because of portability and partially because the main part of PFS not is operational on the Paragon yet.

5. Numerical Experiments

The numerical experiments are divided into two parts. We first consider the performance of the parallel algorithm on the Intel Paragon varying the number of processing nodes and the dimension of the model and then we compare the performance of the full model including I/O on the Intel Paragon, one processor of a

CRAY Y-MP and an IBM RS/6000.

The horizontal resolution of the model is $N_x = 151$ grid points in the longitudal direction and $N_y = 133$ grid points in the latitudal direction. This resolution was used in all the experiments, while tests have been done with both $N_z = 2, 5, 14$ and 20 vertical layers. In all the experiments a period of 6 hours was simulated. As meteorological fields are read four times a day and the time step is 300 seconds, this corresponds to reading one meteorological field, simulating 72 time steps and writing the concentration fields. In order to measure the typical performance of the model, the initialisation, which involves memory allocation and reading of initial data and one meteorological field, is not included in the timing results presented below.

Figure 6 depicts elapsed time for the parallel algorithm (ie. the full model excluding I/O) on the Intel Paragon for four values of N_z with number of processors varying from 1 to 56. The number of subdomains always equals the number of

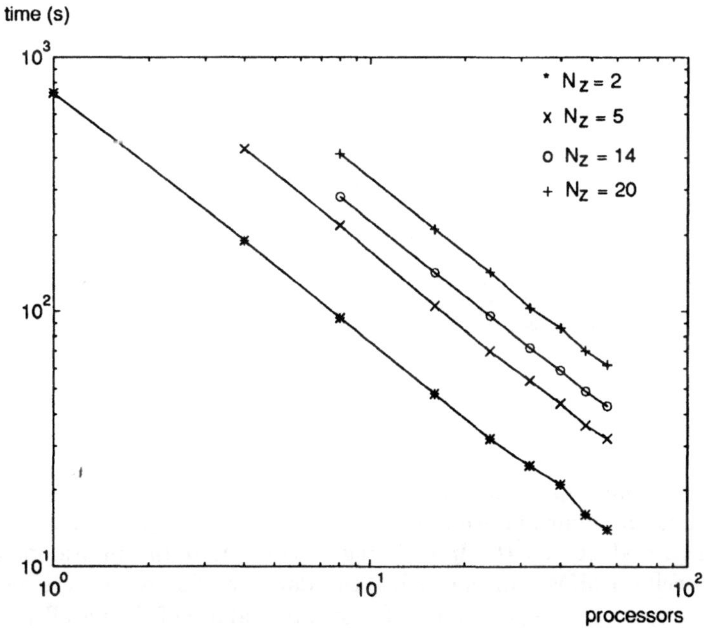

Fig. 6. Elapsed time exclusive I/O as a function of number of processors with four different numbers of vertical layers

processors. The Paragon used in these experiments have 4 times 14 nodes and for all experiments with 4 or more nodes we have therefore used 4 subdomains in the x-direction (longitude) and from 1 to 14 subdomains in the y-direction (latitude). This ensures only neighbour-to-neighbour communication.

Due to memory constraints, the model can only be run on one processor when three or less vertical layers are used. We therefore do not present any speedup results, but we see that the performance of the algorithm scales well for all the experiments. For $N_z = 2$ we have run the single processor version as well as the multiple processor version of the model. The results show an initial drop in efficiency when we go from one to four processors due to the communication. For higher number of processors the decrease in efficiency is small. With four processors the efficiency is 96% and with 56 it is 93%.

We now compare the results for the full model with resolution 151 x 133 x 20 on 56 nodes of the Paragon, one processor on a CRAY Y-MP and an IBM RS/6000 model 580. The elapsed times in seconds and splitting of these into computation, input and output are shown in table I. Both the CRAY and the IBM RS/6000 where run in dedicated mode and had more memory than the total size of the model including data. Also the Paragon was run in dedicated mode, and the available memory on each node is about six times larger than the size of the 56-node version of the model. The compiler and compiler options used on the different machines are:

Intel Paragon: `f77 -nx -O4 -Knoieee`
CRAY Y-MP: `cf77`
IBM RS/6000: `xlf -O3 -Pk -Wp,-r=3,-lc=blas,-ag=a`

TABLE I
Comparison of simulation times for Intel Paragon, CRAY Y-MP and IBM RS/6000

	Elapsed	Computation	Input	Output
Intel Paragon 56 nodes	100	62	21	17
CRAY Y-MP 1 processor	211	195	12	4
IBM RS/6000 model 580	802	766	20	16

6. Conclusion

We have discussed one implementation of a multidomain algorithm for advection-diffusion equations in an air pollution model. The numerical experiments with the parallel model carried out on the Intel Paragon shows that the multidomain algorithm scales well on MIMD machines like the Paragon. The results are also encouraging with respect to the possibilities for operational use of the parallel model. However, the I/O performance needs to be improved. Furthermore, the stability of the system will be essential in an operational environment, as each simulation is expected to take days on a parallel system with the order of 100 processors.

In this first parallel implementation of the algorithm in a real world application we have chosen to parallelise only in the horizontal direction (ie. we have only used the advection part of the algorithm). We have also used the same discretisation scheme for spatial discretisation as in the sequential model. As the multidomain algorithm is a continuous algorithm for full advection-diffusion problems, several other possibilities are available. We plan to do implementations using Chebyshev spectral collocation for space discretisation in the subdomains and a flux balance technique for information exchange between the subdomains as well as parallelisation of the diffusion part of the model. We will also make a PVM-based version of the model, in order to run other parallel systems as well as clusters of workstations. Results from these implementations will be reported later.

Acknowledgements

This work was supported by The Research Council of Norway, department NAVF, grant 447.92/017, by Intel Supercomputer Systems and by SINTEF.

References

[1] A.Bott: *A Positive Definite Advection Scheme Obtained by Nonlinear Renormalization of the Advection Fluxes.* Monthly Weather Review 117(1989) pp. 1006-1015.
[2] A.Bott: *Reply.* Monthly Weather Review 117 (1989) pp. 2633-2636.
[3] R.Esser, R.Knecht: *Intel Paragon XP/S - Architecture and Software Environment.* KFA-ZAM-IB-9305, Forschungszentrum Jülich GmbH, Jülich, Germany, 1993.
[4] S.Grønås, A.Foss, M.Lystad: *Numerical simulation of polar lows in the Norwegian Sea.* Tellus 39A(1987) pp. 334-353.
[5] L.Halpern: *Artificial Boundary Conditions for Incompletely Parabolic Perturbations of Hyperbolic Systems.* SIAM J.Math.Anal. 22(1991) pp. 1256-1283.
[6] I.Lie: *Multidomain Solution of Advection Problems by Chebyshev Spectral Collocation.* Submitted to J.Sci.Comput., 1992.
[7] I.Lie: *Open Boundary Conditions for Incompletely Parabolic Systems.* FFI/REPORT-92/7029, FFI, Kjeller, Norway, 1992.
[8] I.Lie: *Open Boundary Conditions and Interface Conditions for Advection-diffusion Equations.* Draft, Kjeller, Norway, 1993.
[9] I.Lie, R.Skålin: *A Multidomain Algorithm for Advection Problems and its Application to Atmospheric Models.* In Proceedings from Fifth ECMWF Workshop on Use of Parallel Processors in Meteorology, ECMWF, Reading, England, November 23-27, 1992.
[10] M.Macaraeg, C.L.Street: A Spectral Multidomain Technique for Viscous Compressible Reacting Flow. Intern.J.Numer.Meth.Fluids 8(1988) pp.1121-1134.
[11] G.J.McRae, W.R.Goodin, J.H.Seinfeld: *Numerical Solution of the Atmospheric Diffusion Equation for Chemically Reacting Flows.* J.Comp.Phys. 45(1982) pp. 1-42.
[12] T.E.Nordeng: *The effect of vertical and slantwise convection on the simulation of polar lows.* Tellus 39A(1987) pp. 354-375.
[13] *Paragon^{TM} OSF/1 User's Guide.* Revision-001, Intel Supercomputer Systems Division, Beaverton, Oregon, 1993.
[14] H.Sandnes, H.Styve: *Calculated budgets for airborne acidifying components in Europe, 1985, 1987, 1988, 1989, 1990 and 1991.* EMEP/MSC-W, Report 1/92. The Norwegian Institute of Meteorology, Oslo, Norway, 1992.
[15] L.Tarrason: *Contributions to sulphur background deposition over Europe. Results for 1988.* EMEP/MSC-W, Note 5/92. The Norwegian Meteorological Institute, Oslo, Norway, 1992.

MULTIDISCIPLINARY MATHEMATICAL MODEL FOR EARTHQUAKE PREDICTION STUDIES

A.S. Alekseev

Computing Center of the Siberian Division of the Russian Academy of Sciences
Lavrentiev Prospect 6
Novosibirsk 630090
Russia

Abstract

The combined statement of inverse problems for various geophysical fields and use of the solutions obtained for the formation of a complex (multidisciplinary) earthquake precursor are considered. Our aim is the construction of a variant of the multidisciplinary mathematical model describing the formation, propagation and interaction of geophysical fields of different physical nature and changes of their anomalies in the earthquake prone zones for the earthquake prediction studies.

Further developments of the theory and a physical-mathematical model of the multidisciplinary earthquake prediction are considered in the given paper. The solution of the combined inverse geophysical problem for several geophysical methods and anomalies of the corresponding fields measured at the earth's surface in the place of earthquake preparation, as well as the set of geophysical precursors of earthquakes are used as initial data. This model can serve as a basis for the systematization and studies of interactions of precursors of different geophysical nature. It is proved that multidisciplinary earthquake prediction can give some new possibilities for the development of the quantitative method for the prediction of intensity, location and time of a future earthquake by means of the complex monitoring of seismic prone zones using variations of different geophysical fields.

F.-X. Le Dimet (ed.), High Performance Computing in the Geosciences, 187–205.

For the numerical solution of combined inverse problems we suggest and discuss a special technique using high performance parallel computers.

Scientific approach to the prediction of earthquakes is based on three fundamental physical-geological ideas. Preparation and occurrence of earthquakes is through the change of rheological (physical-mechanical and chemical) properties of rocks in the area of future earthquake source; mechanisms of evolutional and catastrophic development of a particular earthquake source are a part of the total seismogenic process; the multi-dimensional and multi-factor seismogenic process is subject to some physical and statistical regularities. Observations of the rheological state of seismic prone areas, analysis of the earthquakes' mechanism and the seismogenic process on the whole can help in the development of scientific methods for the prediction of strength, place and time of occurrence of earthquakes.

Taking into account the complex nature of seismogenic processes, seismologists in different seismically active regions of the earth use in their investigations the whole set of geophysical methods. It has been found out that numerous anomalous changes of different geophysical fields and basic geological and geochemical processes take place before many strong earthquakes. These changes can be considered as precursors of earthquakes.

Over the last decades large material on the complex study of the seismogenic process and precursors has been collected at the Garm geophysical test site by the Institute of the Earth's Physics of the Russian Academy of Sciences. A great number of precursors have been studied. In particular, spatial inhomogeneity of time variations of seismic velocities in the earth-crust was investigated in detail [1], [2].

Significant progress in the prediction of earthquakes was achieved in China. An extensive network of observational stations for the monitoring of seismogenic processes and different precursors (deformation of lithosphere, variations in ground waters, electric conductivity, electromagnetic and gravitational anomalies, etc.) has been developed in the recent 20 years. Complex detailed analysis of about a 100 strong earthquakes that occurred from 1966 to 1990 was carried out and more than a 1000 anomalies that can be considered as precursors were identified. They form a data base of the multidisciplinary investigations. A special function was introduced for the computer prediction — the amount of comprehensive seismic information — and procedures for the prediction of magnitude, place and time of occurrence were developed using this information [3–4]. A number of predictions was successful. In papers [3], [4] it was noted that the works are still at the initial stage and need theoretical and computational development. Earthquakes remain the most dangerous natural catastrophic phenomenon (Figure 1).

Construction of an improved physical-mathematical model of the earthquakes' preparation and occurrence including precursors of different nature as physical elements of this multidisciplinary model is necessary. Extremely important is the development of the 'physical prediction' method [4]. A new and more comprehensive physical-mathematical model allowing the use of up-to-date computational possibilities for the earthquake prediction is necessary.

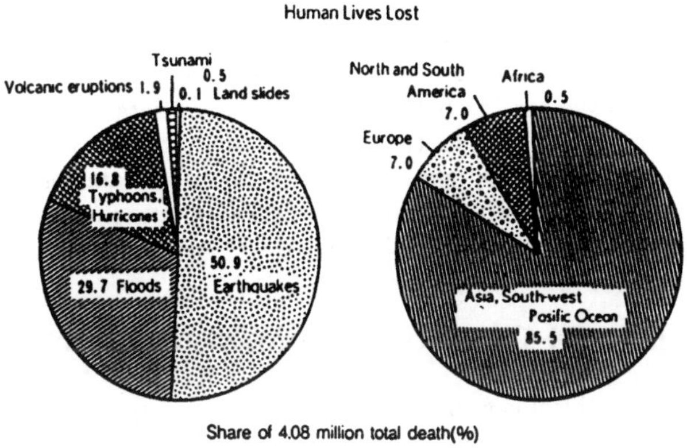

Figure 1

It should be noted that the problem of global and prospecting geophysics similar in its scientific content has become important in the recent years. Here quantitative methods for combining different geophysical methods are being developed. Neither of these methods separately can completely and correctly solve some of the geophysical problems. Here it is also necessary to pass from the existing level where there is an attempt to combine not very reliable results of separate methods into a more reliable 'picture' by means of 'making decisions by ballot', to a new level where several individual methods are organically incorporated into a new multidisciplinary approach with principally new possibilities of integration. It was found out, in particular, that some geophysical methods of different physical nature complement each other when they are applied to one and the same geological object. Due to this separate methods not yielding unique solutions can give unique and stable results when they are combined in one process of data processing [5]. For example, parameters of rock cracking occurring in some area can be simultaneously manifested in anomalies of electric, gravitational and seismic fields, as well as in the regimes of water and gas migrations. Each of the above mentioned anomalies can make its own contribution into the estimate of cracking parameters.

The observed properties show the way to improve the geophysical methods in global and prospecting geophysics [5–8] and also in the problem of earthquake prediction using complex physical-mathematical models that take into account multidisciplinary character of the processes and initial information.

This article attempts to give a formal scheme for the construction of a multidisciplinary physical-mathematical model for the formation and interrelation of geophysical anomalies that can be treated as earthquake precursors. This scheme contains a description of individual anomalies of geophysical fields using the solution of mathematical equations for the corresponding method, and interactions of different factors causing these anomalies as a result of the complex process of the earthquake preparation. The solution of the combined inverse problem for the full set of equations for interacting fields is proposed as a quantitative approach to the earthquake prediction. In this case all the information concerning 'anomalies-precursors' should be used as initial data and the whole set of characteristics describing the evolution of the rheological state of the earthquake-prone zone should be used as the sought-for solution.

1 Individual and Combined Geophysical Methods

An individual direct problem is formulated in the following way: A body D with a boundary S (possibly composite) is given. The geophysical field $U(x,t)$; $x \in R^n$, $t \in R$ (time) occurs in this body naturally or due to external sources. The field $U(x,t)$ reflects the properties of the medium by means of the equation:

$$L(U,\alpha,\beta) = f(x,t), \tag{1}$$

where $L(U,\alpha,\beta)$ is the field operator (special form for the corresponding conservation law), and $f(x,t)$ is the external field source. Boundary conditions at the surface S and initial (usually zero) data are supposed to be given:

$$L(U,\alpha_s,\beta_s)\Big|_S = h(s,t), \qquad s \in S, \tag{2}$$

$$U(x,t)\Big|_{t<0} \equiv 0. \tag{3}$$

The coefficients α and β describe the properties of the geologic medium[1]. There are properties of two types. α-type represents physical characteristics of the medium (density, electrical resistance or elasticity, etc.), and β-type — geometrical characteristics (parameters for the interface of areas filled with media having different physical parameters).

In the appropriate parameterization of the media models, piecewise smooth vectors $\alpha(x,t) = \{\alpha_1(x,t),\ \alpha_2(x,t), ..., \alpha_k(x,t)\}$ and $\beta(x,t) = \{\beta_1(x,t),\ \beta_2(x,t), ..., \beta_r(x,t)\}$ that have one-to-one correspondence to physical and structural properties of the considered

[1]The relations (1)–(3) determine the physical-mathematical model for some geophysical method. Now it is not necessary to fix the physical nature of individual methods in the general scheme for the statement of direct and inverse problems. This will be made at the next stage of the development of the model.

media (specify the model for the medium) can be introduced for the corresponding above indicated two types of properties.

$U(x,t)$ is to be obtained in the given set of observation points in the body D or at its surface S provided $\alpha(x,t)$, $\beta(x,t)$, $f(x,t)$, $h(x,t)$ are specified. Often it is necessary to find not the function $U(x,t)$, i.e. the whole field, but only some functional $F(x_i, t_j)$;

$i = 1, 2, ..., n$, $j = 1, 2, ..., m$ (these are values of the field in the points x_i, t_j or spectral constituents of the field at the given frequency values).

In all the well-known models for the earthquake preparation (creeping of blocks at the surfaces of rupture; dilatancy of the medium; snow avalanche fracturing) and in new models (for example, the model for the volume deformational reconstruction [9]), asymmetry of stressed or deformed state is always present. This must create anisotropy of the medium mechanical properties. The extent of anisotropy (seismic coefficient of anisotropy) can be directly related to the earthquake preparation phase. If the existence of such relation is assumed then a quantitative parameter appears describing the dynamics of the rheological state of the earthquake-prone zone by means of the recording of new seismic waves and their anomalous polarization forming in anisotropic media. Amplitudes and frequency spectra of these waves are directly related to the coefficients of the medium anisotropy and contain information on the process of the earthquake preparation.

In the direct problems of geophysics, the coefficients $\alpha(x,t,U)$ and $\beta(x,t,U)$ should be given from a priori considerations, resulting from concrete studies. For the problems of earthquake prediction this question has some features that will be pointed out below in connection with the discussion of possibilities for vibroseismic sounding of a seismic prone zone and seismotomographic monitoring of rheological anomalies in it.

Numerical methods for the solution of geophysical direct problems are well-developed now [10–11]. However, in the case of non-linear dependence of coefficients $\alpha(x,t,U)$, $\beta(x,t,U)$ on the solution $U = U(x,t)$ this question can be very complicated. In these cases, we can try to linearize the equations (1) for direct problems.

The inverse problem for an individual geophysical method is formulated in the following way: it is necessary to find the coefficients $\alpha(x,t)$ and $\beta(x,t)$ from the equation set (1)–(3) according to the given geophysical field

$$U(x,t)\Big|_S = U^0(s,t) \tag{4}$$

at the surface of observation S. The function $U^0(s,t)$ is known from measurements. Besides $\alpha(x,t)$ and $\beta(x,t)$, the volumetric field sources $f(x,t)$ and the boundary conditions $h(s,t)$ from (2) sometimes have to be among the sought-for elements.

Questions of the correctness of statement of the inverse geophysical problem, i.e. existence, uniqueness and stability of their solution, were studied for individual geophysical methods: see P.S. Novikov [12] (gravitational method), A.N. Tikhonov [13] (electrical method), M.M. Lavrentiev [14] (potential fields), A.S. Alekseev [15] (nonstationary seismic fields), V.G. Romanov [16] (nonstationary geoelectric fields). At present a number of numerical methods for the solution of inverse problems was developed that make it possible to obtain their effective solution in typical situations.

Now consider a situation when for one and the same geological object, several geophysical fields generated by this object can be observed. It is necessary to determine all its physical (petrophysical), rheological and geometrical characteristics. Then the following problem arises:

Each field $U_\nu(x,t)$ ($\nu = 1, 2, ..., m$, where m is the number of physically individual fields) is described by its mathematical equation:

$$L_\nu(U_\nu, \alpha_\nu, \beta_\nu) = f_\nu(x,t), \tag{1^ν}$$

Physical and geometrical features of the body $\alpha_\nu(x,t)$ and $\beta_\nu(x,t)$ are assumed to be piecewise-smooth functions of the coordinates $x = \{x_1,\ x_2, ..., x_n\}$. These functions have a finite number of discontinuities which may be different for different ν.

There are boundary conditions and initial data for each method:

$$l_\nu(U_\nu, \alpha_\nu, \beta_\nu)\Big|_S = h_\nu(s,t), \tag{2^ν}$$

$$U_\nu(x,t)\Big|_{t<0} \equiv 0. \tag{3^ν}$$

The totality

$$U_\nu\Big|_S = U_\nu^0(s,t),\ \nu = 1, 2, ..., m \tag{4^ν}$$

of different fields at the body surface is given as initial information.

Here $U_\nu^0(s,t)$ are known functions. It is necessary to find

$$\alpha_\nu(x,t) \cdot \beta_\nu(x,t). \tag{5}$$

Definition. An inverse problem is called a combined inverse problem of several geophysical methods, if the whole equation set (1^ν) at $\nu = 1, 2, ..., m$ is considered as equations for the generalized geophysical process, the equation sets (2^ν) and (3^ν) are taken as generalized boundary conditions and initial data, and the whole totality of m fields $U_\nu(s,t)$ measured at the surface S of the considered body is taken as initial information.

It is easy to note the following important property of the combined inverse problem: it is not equivalent to the totality m of separate (independent) problems.

In fact, if the solution of each of the individual inverse problem (1^ν)–(4^ν) is unique at all ν, then the combined inverse problem has the property of solution uniqueness (its solution must consist of the totality of solutions $\alpha_\nu(x,t)$ and $\beta_\nu(x,t)$ of all the individual problems). But the inverse statement in the general case is not correct. If some of geometrical characteristics of the body $\beta_\nu(x,t)$ are identical for different methods with the numbers ν_1 andν_2, then the combined inverse problem has less 'degrees of freedom' and its solution can be unique, even if any of the functions $\beta_1(x,t)$ or $\beta_2(x,t)$ in one of the individual inverse problems cannot be uniquely defined. The case of an isolated homogeneous inclusion with the boundary S, for which different inverse problems (gravitational, seismic, etc.) are formulated, can be a particular example of this situation. Since the body is each time the same, the shape of the boundary S can be determined by only one method (seismic, for example). Then the information deficit of the other method (gravitational method in our case) can be compensated by the boundary S known from seismic data. The combined inverse problem is characterized by this mechanism of data deficit compensation in one method at the account of information redundancy in other methods.

The fact of nonequivalence between the combined problem and individual problems rather important for applications deserves more detailed consideration, by presenting a number of particular examples different in their physical and mathematical content.

In the case of the combined inverse problems there arises a number of important questions associated with the choice of particular methods, the amount of initial data, ways of parameterization of models, that would provide the correct solution of the geological-geophysical problem. Therefore it is important to study the basic characteristics of the combined problems, such as existence, uniqueness, stability, and to develop correct numerical methods for their solution.

Without the study of these questions, practical ways of combining different geophysical methods are fairly subjective and are similar to the skill of synthesizing an abstract picture from the most impressive elements of pictures produced by individual methods independently.

Now let us give some general properties of combined inverse problems that either can be easily deduced from their non-equivalence to totalities of individual problems or can be considered as problems for special analysis.

Non-equivalence of problems associated with the uniqueness of their solutions is mentioned above. Here, as it is easily seen, the following variants exist:

1) *According to the fact of existence:* if the solutions of all the individual problems exist, the combined problem can have no solution (variant 1.1), if there is no agreement between the separate fields (4^ν) at the surface (any pair of fields from the set $U_\nu^0(s,t)$); or it can have a solution (variant 1.2), if there are such agreements (all the $U_\nu^0(s,t)$ are mesured and given correctly).

2) *According to the fact of uniqueness:* if the solutions of all the individual problems are unique, then the solution of the combined problem is unique (variant 2.1). If the solution of any individual problem is non-unique, then the solution of the combined problem is either unique (variant 2.2), or non-unique (variant 2.3).

In the variant 2.2, the combined problem can have a unique solution also in the case when several individual problems or even all of them do not have uniqueness (such examples are given below). This property is important and deserves special analysis.

3) *Consideration of variants of relation between problems according to the fact of the solution stability* is interesting. But this requires special study since the character of stability is usually determined analytically with the help of several numerical parameters, and detailed consideration, that is impossible in the given article, is necessary for the selection of 'discrete' variants.

Ending our classification of variants of interrelation between the fundamental properties of individual and combined inverse geophysical problems, it is necessary to point out the main property of the combined problems that can be called the property of *geophysical complementability*. It is in the following. If there is non-uniqueness and instability of solutions of individual inverse geophysical problems, the solution to the combined problem can be unique and stable. Individual methods help in the process of solving the general geophysical problem by means of all their information resources [27].

This property of complementability can play an important role in the optimization of complex geophysical investigations improving their quality and reducing expenses for field observations.

It can turn out to be useful for the development of multidisciplinary methods of earthquake prediction, helping to use geophysical fields of different physical nature for different purposes, to identify interrelations between their anomalies and to use these interrelations in the quantitative form.

Examples of concrete combined problems will be given below. They show that the effect of complementability and multi-disciplinarity increases when any complementary relations between the sought-for coefficients of the medium, fields or their anomalous values are used.

2 Classes and Examples of Combined Inverse Problems in Geophysics

As a result of analysis of the statement of combined problems, several types of such problems can be singled out. They differ in the physical and/or geometrical character of additional relations between the elements of mathematical models for these problems.

Let us consider the following cases of global and prospecting geophysics. The problem of earthquake prediction can evidently have both the variants of these cases and new types.

A. Some geometrical characteristics of a geological or geophysical object turn out to be the same relative to two or more geophysical methods. Isolated homogeneous inclusion with the boundary S studied by different methods is an example of this.

B. Some coefficients of equations for different geophysical methods are identical (they have similar physical meaning and/or quantitative form). Density in the equations for the propagation of seismic waves and in gravics equations is an example.

C. Some coefficients included into different equations of the corresponding methods have functional or statistical connection.

The basic problem of petrophysics, i.e. the determination of the geological type of studied rocks using their geophysical fields and physical properties, is an important example for this case. It should be pointed out that the combined statement of inverse problems for all the existing geophysical methods is a real variant of the mathematical formulation for the inverse problem of petrophysics.

A large volume of information on statistical relations between individual manifestations of the seismogenic process has been accumulated in seismology. At present this information is quantitatively used only in statistical methods of analysis and prediction and, in fact, independently from the physics of processes.

Nevertheless, the structure of the considered type of combined inverse problems completely corresponds to the tendency of quantitative use of statistical relations between fields, anomalies, and properties of rocks in geophysics.

In our opinion, one example of problem statement of this type that will be presented below gives a variant of quantitative physical model for multidisciplinary studies in the problems of earthquake prediction. Here a possibility for the combination of multidisciplinary methods discussed in [4] and the proposed model for the formulation of combined inverse problems is easily seen.

D. Mathematical equations for the geophysical process have coefficients characterizing different physical fields. For example, there exist equations for seismo-electromagnetic processes including the displacement vector of the body points and two vectors of the electromagnetic field caused by the electroconducting rocks, vibrating in the Earth's magnetic field [17]. Here all the three fields should be used for the solution of the inverse problem. The problem can be assigned to the class of combined problems, even if only one mathematical equation is considered as the law of the fields interaction. A similar situation is in the seismo-gravitational equation of A. Love [18], where the gravitational effect of large masses of rocks that are being compressed and oscillate with low spatial frequencies, is taken into account.

These effects are important for the investigation of vibrations of the Earth itself and of larger planets. Such problems will be called combined problems in the sense that several fields interact.

The given classification of the combined problems is not complete. We need it here to illustrate how large is the class of these problems.

The first known to the author statement of the combined geophysical problem of the A-type belongs to I.P. Nedialkov, Bulgarian geophysicist [19], who studied the problem of the quantitative combination of methods associated with potential fields. In [19–20], he shows that the combined interpretation of anomalies of two fields, gravitational and electrical, for example, generated by one and the same isolated geological object, can give more than just a sum of results of individual consideration of each anomaly. The most illustrative classical example given by him is the interpretation of two anomalies of the sphere or cylinder. It is well-known that the location of the center, the radius and redundant density of a cylinder (or a sphere) cannot be uniquely defined from the gravitational anomaly measured on the plane not intersecting the cylinder. It is possible only to construct a two-parameter family of cylinders, whose depth of the center, radius and density are connected by one functional relation.

Similarly, only a two-parameter family of cylinders, whose depth of the center, radius and redundant electric conductivity are connected by one relation, can be determined from the electrical field of the cylinder.

Therefore, having two separate (gravitational and electrical) methods, we are not able to solve the inverse problem of geophysics. None of the methods gives a unique solution. As a result, we cannot determine the depth of the cylinder's center, radius, density and electric conductivity (the four desired parameters). However, remembering that our two anomalies refer to one and the same geological object and setting the same geometrical characteristics in the two (gravitational and electrical) families, one can uniquely define all the four parameters. I.P. Nedialkov has studied a wide class of problems related to the combination of potential fields for the bodies with complex geometries.

Problems of the B-type and their properties can be demonstrated on the papers [6], [7] as an example, which consider the model statement of the combined inverse problem of seismoacoustics and gravics.

Assume [6], that the modulus of volumetric elasticity of the medium $\lambda = \lambda(x, y, z)$ is a smooth function, the shear modulus is $\mu = 0$, the density of the medium $\rho = \rho(x, y, z)$ is also a smooth function. The medium fills the half-space $z > 0$ of the $3D$ space.

Processes of wave propagation are described by the system of equations:

$$L\vec{U} = \rho(x, y, z)\vec{U}_{tt} - grad[\lambda(x, y, z) div\vec{U}] = 0, \quad t \geq 0, \ z > 0, \tag{6}$$

for the initial data

$$\vec{U}\Big|_{t=0} = U_0(x, y, z), \quad \vec{U}_t\Big|_{t=0} = 0, \tag{7}$$

and the boundary conditions

$$T(\vec{U})\Big|_{z=0} = 0, \tag{8}$$

where T is the stress operator on the plane containing the first derivatives of $\vec{U}$ along the coordinates.

For solving the inverse problem we set, additionally, the condition:

$$\vec{U}\Big|_{z=0} = \vec{U}_1(x,y,0,t). \tag{9}$$

Our medium is in the gravity field and generates the gravitational field. The gravity field potential $\Phi(x,y,z)$ is determined from the equation

$$\Delta\Phi = -4\pi\rho, \quad z > 0. \tag{10}$$

In the inverse problem, the potential is assumed to be given on the plane $z = O$ by the relation

$$\Phi\Big|_{z=0} = \Phi_0(x,y). \tag{11}$$

As the combined inverse problem for the system (6)–(11) we understand the problem of determining the functions $\rho(x,y,z)$ and $\lambda(x,y,z)$ from the given $\vec{U}_0$, $\vec{U}_1$ and Φ_0. If we succeed in determining p and λ uniquely, it will be possible to find $\vec{U}(x,y,z,t)$ and $\Phi(x,y,z)$ by solving the corresponding direct problems.

Note that the individual problem (10)–(11) for the determination of density does not have a unique solution without strict a priori assumptions on the properties of the sought-for function. Similarly, it is impossible to obtain a unique solution of the inverse problem (6)–(9) without strict assumptions on the functions $\lambda(x,y,z)$ and $\rho(x,y,z)$.

In [6], the theorem of uniqueness of the combined problem has been obtained under essentially less strict restrictions imposed on the class of the sought-for functions. It is only necessary to have sufficient smoothness and boundedness of functions and their derivatives up to the third order in R^3, and also their factorization of the type $\lambda(x,y,z) = \lambda_1(x,y)\lambda_2(z)$.

In [7], stability of the solution of the combined problem (6)–(11) is shown (and the estimate is given) under the conditions when individual problems have no unique solutions.

As an example of problems of the C-type (see [8]), let us consider the combined problem for m arbitrary geophysical methods described by the equations

$$L_\nu(U_\nu, \alpha_\nu, \beta_\nu) = f_\nu(x,t) \qquad \nu = 1, 2, ..., m. \tag{12}$$

For each field U_ν, as in the problem (1^ν)–(5^ν), we have the boundary conditions and initial data which are necessary in inverse problems.

Assume that there are statistical connections of the correlational type (as is assumed in petrophysics [21]) between different physical properties α_ν of the considered object.

Let all the vectors $\alpha_\nu = \{\alpha_{\nu_1},\ \alpha_{\nu_2},...,\alpha_{\nu_k}\}$ and $\beta_\nu = \{\beta_{\nu_1},\ \beta_{\nu_2},...,\beta_{\nu_r}\}$ be combined in one vector $\alpha = \{\alpha_{11},\ \alpha_{12},\ \alpha_{21},\ \alpha_{22},...,\alpha_r\}$, denoted as $\alpha = \{\alpha_1,\ \alpha_2,...,\alpha_L\}$.

Assume that the components of the vector are random functions with the known values $< \alpha_i >= \alpha_i^0$. The total vector $\alpha(x)$ has the known mean value $\alpha^0(x)$ and the known covariance matrix $G_a(x)$. A priori information about these functions can be obtained for the certain parts of the Earth crust from petrophysical reference-books and geological reports. In seismology, the mean values of α_i^0 can be obtained from earlier studies or from special observational systems for the continuous monitoring of geophysical parameters.

A possibility to add to the system (12) new equations, that are not necessarily differential, but functional or correlational relations between parameters and fields, if such relations exist a priori (in a certain data base), or if they appear in the course of monitoring, is an important property of the combined inverse problems.

Let direct problems for each equation (12) with the corresponding boundary and initial conditions have the unique solutions $U_\nu(x,t)$, $\nu = 1, 2, ..., m$ for the sufficiently smooth $\alpha(x)$.

Our combined inverse problem is in the determination of random functions $\alpha(x)$ and in the estimation of their variances from the initial information given in the form:

$$V_\nu(\bar{x},t) = U_\nu(\bar{x},t) + \epsilon_\nu(\bar{x},t). \tag{13}$$

Here $\bar{x} \in S_\nu \subset D$, D is the geological object in question, S is the set of points, where the fields $U_\nu(\bar{x},t)$ are measured. The vector $\epsilon(\bar{x},t) = \{\epsilon_\nu(\bar{x},t)\}$ characterizing measurement errors is assumed to be normally distributed with the covariance matrix $G_\epsilon(\bar{x},t)$.

The following results can be obtained.

The accuracy (variance estimate) and stability of the solution can be characterized by the following relations:

$$\sigma_l^2 \leq 2^{2(1-m)} \max_{\nu=1,m} \{\sigma_{\nu l}\}\sigma_l^0; \quad l = 1, 2, ..., L, \tag{14}$$

$$\xi \leq \max\{\zeta,\phi\}, \quad \phi \leq \max_{\nu=1,m} \{\xi_\nu\}, \tag{15}$$

where: σ_l is variance in the determination of $\alpha_l(x)$ in the combined problem; $\sigma_{\nu l}$ is variance in the determination of $\alpha_l(x)$ in the individual method with the number ν (the number l in our succesive enumeration defines the corresponding number ν and the order of the coefficient in the corresponding equation); σ_l^0 is the square root from the diagonal element with the index (l,l) in the matrix $G_{\alpha^0}^{-1}$.

For the relations (15) determining the solution stability, we introduce the following notations: ξ is the condition number of the Fisher matrix in the combined statement; ζ is the condition number of the matrix $G_{\alpha^0}^{-1}$; ζ_ν is the condition number of the Fisher matrix in the statement of the individual method with the number ν, Ψ is an auxiliary quantity.

With the increase in the number of combined methods in accordance with the relation (14), the accuracy of determining the coefficients $\alpha(x)$ on the whole increases (variance decreases). However, the addition of a new method with a fairly large variance $\sigma_{\nu l}$ can overpower the effect of accuracy increase accumulated in the combination of the previous methods. Such methods with sharp peaks of variances must be excluded.

Inverse problems of seismics and geoelectrodynamics can be taken as the specific combined problems of C-type considered above (with the statistical connection between the components) using the correlations between the seismic and electromagnetic characteristics of rocks in the given area, investigated with the help of the corresponding monitoring systems [22].

It was shown that the growth of effectiveness of geophysical methods with the increase of the number of separate methods that are being combined (in the quantitative sense) is due to the fact that the growth of the number of unknown parameters is not as quick as the increase of the volume of additional initial information.

The mechanism mentioned above can be more effective if we pass in the classical models of geophysical methods from the phenomenological level of averaging to the next level, i.e. microphysical. This can be done by introducing geologically-oriented structural-physical parameters (porosity, skeleton elasticity, fluid mobility, electromagnetic properties of the fluid and skeleton, etc.) into models, instead of the averaged parameters having the phenomenological sense (λ, μ, ρ in seismics, σ, μ, ϵ in electrodynamics, etc). This will immediately lead to coincidence of a larger number of coefficients, included into the equations of various geophysical methods.

3 Computational Aspects

The numerical solution of combined inverse problems both in geophysical prospecting and earthquake prediction requires development of special numerical methods and the use of powerful computers. The numerical approach is based on the solution of the direct problems for individual methods (now they are applied to rather general geophysical models) and optimization methods for the misfit functionals in the solutions of direct problems with the measured fields [23–24].

In the numerical solution of combined problems using the optimization method, their basic properties (complementability, stability increase, etc.) are demonstrated in the structure of the sum functional and require special study, like all the numerical optimization methods for complex functionals on the whole.

The most common method used for the numerical solution of combined inverse problems is combined inversion using the optimization of the sum misfit functional

$$F(\alpha,\beta) = \sum_{\nu=1}^{m} \lambda_\nu \sum_{s=1,k=1}^{n_\nu,r_k} \left[U_\nu(x_s,t_k) - \hat{U}_\nu(x_s,t_k,\alpha_\nu,\beta_\nu)\right]^2, \tag{16}$$

where x_s, t_k are observation points of geophysical processes $U_\nu(x,t)$; $\hat{U}_\nu$ are variants of geophysical fields that are being computed for the changing medium models (α_ν, β_ν); λ_ν are weight coefficients of the methods.

One of the most essential computational problems of optimization is the necessity of quick solution of direct problems (computations $\hat{U}_\nu$) for a large number of variants of models. Solution of problems of propagation of elastic seismic waves in extensive areas of the earth crust, in large frequency intervals and wave propagation times is computationally very expensive.

As we don't have 'supercomputers', we had to develop a kind of 'superalgorithms'. One of such algorithms utilizes analytical separation of variables by the Fourier or Fourier-Bessel trasnforms or Legendre series in order to split $3D$ elastic-dynamic problems for inhomogeneous media into a series of $1D$ problems and then solve the latter one after another with the help of the $1D$ finite-difference technique. This algorithm is 8–10 times more time-effective than conventional numerical methods.

The main advantage is that the algorithm can be easily applied to parallel computing. Essentially, for each spatial frequency the components of a vector field can be computed on an individual processor, as for the series of $1D$ problems. After this parallel calculations are summed in a host processor. Thus, increase in the number of processors gives the linear increase of performance of our algorithm.

An example of this type of algorithm is the solution of the $3D$ dynamic problem for anisotropic inhomogeneous half-space [11].

The parallel computational system 'Siberia' on the basis of 3 computers EC-1066 (of the IBM-main frames type), 8 array processors EC-2706 (of the type AP-190 L) and special software (a parallel translator of consecutive programs from Fortran) were developed in the Computing Center of SDRAS in addition to the quick algorithms for the solution of direct problems for real tackling of complex questions including optimization of functionals of the type (16). This computational system is shown in Figure 2. Its performance for 4 types of problems is given in Figure 3. It should be noted, that this computer is not powerful enough for the solution of real problems of earthquake prediction.

The concept of multidisciplinary monitoring of seismic prone zones is organically combined with the concept of deep analysis of physical nature of geophysical processes using the statement and solution of the combined inverse problems. A possibility to add new differential equations (new methods) and relations with the statistical or deterministic connection between the fields and parameters of the medium, to the system (1^ν)–(4^ν), in the process of analysis (i.e., "in the process of mathematical monitoring") plays an important role here. In this connection the concept of multidisciplinary investigations becomes more physical and dynamical in this updating process.

In the problems of earthquake prediction this concept has large scientific potential for the highly productive computers, both for the numerical modelling of different geophysical processes, intellectualization of interpretation methods, and operative processing of real data during the "com)utational monitoring" of the earthquake preparation process.

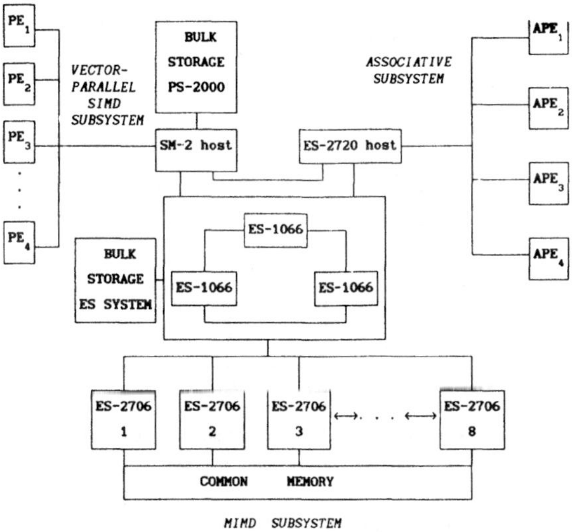

ARCHITECTURE OF MULTI PROCESSOR SYSTEM SIBERIA

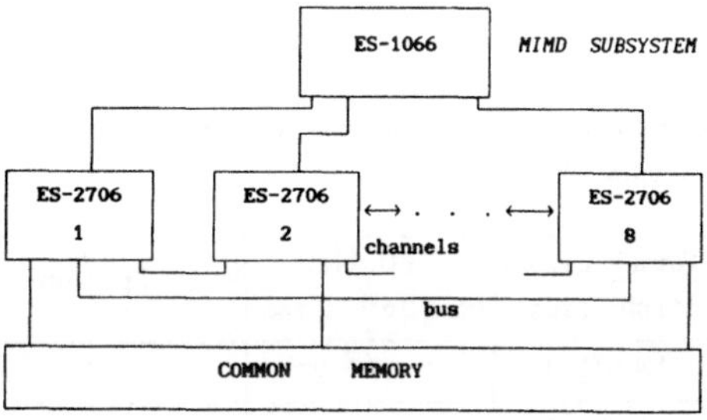

SOFTWARE SUPPORTED COMMUNICATIONS

Figure 2

SIBERIA MULTICOMPUTER SYSTEM PERFORMANCE

1. Frequency filtration for 880×1310 data array (an image)

Computer system	Total elapse time (sec)	CPU time ES-1066 (sec)	Speedup
ES-1066	1509	1509.00	1.0
ES-1066 + 1 PE	361	6.17	4.18
ES-1066 + 2 PE	207	6.19	7.28
ES-1066 + 4 PE	115	6.35	13.16

2. Quantum chaos problem

Computer system	Total elapse time (sec)	CPU time ES-1066 (sec)	Speedup
ES-1066	604.0	604.0	1
ES-1066 + 1 PE	25.0	0.3	24
ES-1066 + 2 PE	12.6	0.3	48
ES-1066 + 4 PE	6.5	0.4	93

3. Spectral and spectral-finite difference methods for computing the synthetic seismograms for 2-D inhomogeneous media.

Computer system	Total elapse time (sec)	CPU time ES-1066 (sec)	Speedup
ES-1066	240.0	240.0	1
ES-1066 + 1 PE	48.0	6.1	5

4. Spectral (spectral-finite difference) method for computing the synthetic seismograms and snapshots.

Computer system	Total elapse time (sec)	CPU time ES-1066 (sec)	Speedup
ES-1066 + 1 PE	14421.0	61.0	1
ES-1066 + 4 PE	5484.0	68.0	2.7

Figure 3

References

[1] Kondratenko A.M., Nersesov I.O.,1962, Some investigation results of velocity change of longitudinal waves and relations between velocities of longitudinal and transverse waves in the earthquake-prone zone, Physics of Earthquakes and Seismics of Explosions, Moscow, Izd. Akad. Nauk SSSR, pp.57–94 (in Russian).

[2] Nersesov I.L., Popandopulo G.A.,1988, Spatial inhomogeneity of temporal variations of velocity parameters in the earth crust of the Garm region, Izv. Akad. Nauk SSSR, ser. Fizika Zemli, No. 8, pp. 13–24 (in Russian).

[3] Mei Shirong, 1992, Progress in earthquake prediction in China during the 80'ies, Journal of Earthquake Prediction, No. 1, pp. 43–57.

[4] Zhang Guomin, Zhang Zhaocheng, 1992, The study of multidisciplinary earthquake prediction in China, Journal of Earthquake Prediction, No. 1, pp. 71–85.

[5] Alekseev A.S., 1992, On the statement of combined inverse geophysical problems, Conditionally Correct Problems of Mathematical Physics and Analysis, Nauka, Novosibirsk, pp. 7–18 (in Russian).

[6] Alekseev A.S., Bubnov B.A., 1981, On one combined statement of inverse problems of seismics and gravimetry, Dokl. Akad. Nauk SSSR, vol. 251, No. 5, pp.1086–1090 (in Russian).

[7] Alekseev A.S., Bubnov B.A.,1984, Solution stability for the inverse problem of combining seismics and gravimetry, Dokl. Akad. Nauk SSSR, vol. 275, No. 2, pp. 332–335 (in Russian).

[8] Alekseev A.S., Erokhin G.N., 1989, Cooperative approach to solving inverse problems in integral geophysics (theoretical aspects), Geophysical Data Inversion Methods and Application, Proceedings of the 7th International Seminar, Free University of Berlin.

[9] Vostrikov Yu.N.,1989, On the physical basis of seismology of earthquake-prone zones (the state and hypothesis of deformational transition). Preprint No. 18. Institute of Geology and Geophysics, Siberian Division of Russian Academy of Sciences, Novosibirsk (in Russian).

[10] Alekseev A.S., Mikhailenko B.G., 1974, On the Lamb problem for inhomogeneous semispace, Dokl. Akad. Nauk SSSR, vol. 214, No. 1, pp. 84–86 (in Russian).

[11] Martynov V.N., Mikhailenko B.G., 1988, Calculation of complete theoretical seismograms for anisotropic media. Inverse Modeling in Exploration Geophysics. Proceedings of the 6th International Mathematical Geophysics Seminar. Free University of Berlin, Feb. 3–6, 1988 , Friedr.Vieweg and Sohn Braunschweig/Wiesbaden, pp. 323–348.

[12] Novikov P.S., 1938, On the inverse problem uniqueness of the theory of potential. Dokl. Akad. Nauk SSSR, No.19, pp. 57–60 (in Russian).

[13] Tikhonov A.N., 1950, On the definition of electrical characteristics of deep earth crust layers, Dokl. Akad. Nauk SSSR, vol. 2, pp. 295–298 (in Russian).

[14] Lavrentiev M.M., 1962, On some ill-posed problems of mathematical physics, Nauka, Novosibirsk (in Russian).

[15] Alekseev A.S., 1962, Some inverse problems in the theory of wave propagation, I, II, Izv. Akad. Nauk SSSR, ser. Geofiz., No. 11, pp. 1514–1531 (in Russian).

[16] Romanov V.G., 1982, Inverse problems of seismic and electromagnetic wave propagation. Methods for the solution of ill-posed problems and their applications. Novosibirsk, Computing Center of Siberian Division of Academy of Sciences of USSR, pp. 111–118 (in Russian).

[17] Novitsky V., 1986, Electromagnetic effects in solid bodies, M, Mir.

[18] Love A., 1911, Some problems of Geodynamics, Cambridge.

[19] Nedialkov I.P., 1957, Complex interpretation of potential fields, Dokl. BAN, vol. 10, No. 6, pp. 67–70 (in Russian).

[20] Nedialkov I.P., 1965, Complex interpretation of potential fields, Izv. Akad. Nauk SSSR, ser. Fizika Zemli, No. 11, pp. 48 – 65 (in Russian).

[21] Lebedev T.S., Korchin V.A., Savenko B.Ya. and others, 1988, Petrophysical investigations under high (p–T) conditions and their geophysical applications, Kiev, Nauk. Dumka (in Russian).

[22] Gokhberg M.B., Gufeld I.L., Marenko V.F. and Ponomarev E.A., 1989, Study of seismic influence on the ionosphere by super, long wave probing of the Earth- ionosphere waveguide, Physics of the Earth and Planetary Interios, Vol. 576, pp. 64–67.

[23] Golizdra G. Ya., 1975, On complex interpretation of seismic and gravitational fields. Dokl. AN Ukr. SSR, ser. BT, vol. 12, pp.114–117 (in Russian).

[24] Lines L., Schultz A., Treitel S., 1988, Cooperative inversion of Geophysical Data. Geophysics, No. 53 (1), pp.120–135.

[25] Biot M.A., 1956, Theory of propagation of elastic waves in a fluid-saturated porous solid.I, Acoustic Soc. America, vol. 28, pp. 168–178.

[26] Genadinnik B.I., Melnikov V.P., Genadinnik G.V., 1986, Theory of induced electrochemical activity of rocks, Yakutsk (in Russian).

[27] Alekseev A.S., 1992, Complementary features of geophysical methods and computational aspects of joint data inversion, 54th Meeting of European Association of Exploration Geophysicists, Paris, pp. 750–751.

Basic functional representation of programs for automatic differentiation in the *Odyssée* system

N. Rostaing-Schmidt, and E. Hassold
Projet SAFIR
Université de Nice-Sophia Antipolis
Laboratoire de mathématiques URA 168 du CNRS
Parc Valrose
BP 71
06108 Nice Cedex 2

ABSTRACT. We present a functional representation of a certain class of *fortran* programs based on the mathematical notions of tangent fiber bundles and cotangent fiber bundles. We deduce from this formalism the rules to apply to produce automatically the tangent linear code and the adjoint code by manipulating the initial program. We present the *Odyssée* system and its application to two meteorological projects: *arpege/ifs* and *arps*. We also describe how we build and analyse a program dependance graph to automatically deduce some interprocedural information needed by the automatic differentiation process.

1. Introduction

Methods to build a program computing the derivatives (partial, in a direction, of different orders ...) of a mathematical function given by a program are called automatic differentiation techniques. Among the existing systems, we can cite ADOL-C [Griewank, Juedes et Srivasan, 1990], ADIFOR [Bischof et al., 1992] or AMC [Giering, 1993]. In this paper, we outline our approach based on computer algebra and program transformation techniques.

In order to master the operations performed during program transformations devoted to automatic differentiation, we have developped a formalism based on the mathematical notions of tangent fiber bundles and cotangent fiber bundles, applied to a composition of functions given by a certain class of FORTRAN programs. We consider the class of ***piecewise differentiable functions*** and present the corresponding class of programs.

We have developped a system, *Odyssée*, which performs automatic differentiation on FORTRAN programs by applying rules deduced from the previously mentionned formalism. We present this system and its application to two very different meteorological projects. The first one, ARPEGE/IFS, is a short and middle range weather forecasting project, developped at Météo-France and ECMWF [Courtier et al., 1991]. We have produced the tangent linear and the adjoint codes [Le Dimet et Talagrand, 1986], according to the strategy used in this application [Rostaing, Dalmas and Galligo, 1993]. The second application we treat in collaboration with F.X. Le Dimet's group at LMC (Université J. Fourier Grenoble and INRIA RHONE ALPES) is the ARPS [CAPS, 1992] project developped in the K. Droegenmeier team at the University of Oklahoma. ARPS aims to predict storms, then short term forecasts are performed in a local area. The ARPS application allows to study other strategies and to test tools to deduce automatically some interprocedural information instead of systematically asking the user to fill the system with. For example, the variables with respect to with a subroutine is differentiated may be deduced from the dependance graph of the global program.

The first part of this paper describes the functional representation of a FORTRAN program we have developped and the rules we can deduce to produce the tangent linear code and the adjoint code corresponding to a given program. The next part concerns the *Odyssée* system. We briefly present the system and its application to the ARPEGE/IFS programs. We describe more precisely the extensions developped to treat a

F.-X. Le Dimet (ed.), High Performance Computing in the Geosciences, 207–222.

global code and their application to the ARPS code. At the end of this paper, we draw some conclusion and present perspectives of this work.

2. Functional representation of a FORTRAN program

It is possible to associate to each assignment $t := f(x, y)$, a function $t = f(x, y)$ and a differentiable mapping such that:

$$\begin{array}{rcl} \tilde{f} : D_{n+p+s} & \longrightarrow & D_{n+p+1} \\ (x, y, z) & \longmapsto & (x, y, f(y, z) = t) \end{array}$$

where each vector x, y, z, of size n, p, s represents respectively:

- the non used variables in this computation but kept in memory
- the variables used and kept
- the variables used and eliminated.

The identifier t is the name given to the obtained formula.

With this representation, a sequence of statements becomes a composition of functions. In order to describe accurately the treatment performed on the statements of a FORTRAN program, we introduce some mathematical notions.

2.1. Mathematical notions

2.1.1. Tangent fiber bundle

DEFINITION 2.1. *Let D_n be an open set of $\mathbb{R}^n$.*
$TD_n = D_n \times \mathbb{R}^n$ is a fiber bundle on D_n called **tangent fiber bundle**

We associate to a differentiable mapping f_1 from an open set D_{n_1} to an open set D_{n_2} a mapping, denoted Tf_1, from TD_{n_1} to TD_{n_2}.

DEFINITION 2.2.

$$\begin{array}{ccc} D_{n_1} \times \mathbb{R}^{n_1} & \xrightarrow{Tf_1} & D_{n_2} \times \mathbb{R}^{n_2} \\ (x, \mathbf{v}) & \longmapsto & (f_1(x), T_x f_1.\mathbf{v}) \\ \downarrow & & \downarrow \\ D_{n_1} & \xrightarrow{f_1} & D_{n_2} \end{array}$$

In the canonical basis of vector spaces $\mathbb{R}^{n_1}$ and $\mathbb{R}^{n_2}$ the matrix of $T_x f_1$ is the jacobian matrix of f_1 with respect to x: $(\frac{\partial f_{1_i}}{\partial x_j})_{ij}$.
Tf_1 is called the **tangent linear application.**

LEMMA 2.1. *Tangent linear application of a composed function: if $f = f_2 \circ f_1$ then $Tf = Tf_2 \circ Tf_1$*

2.1.2. Cotangent fiber bundle

In the following, $(\mathbb{R}^n)^\star$ denotes the set of linear forms from $\mathbb{R}^n$ to $\mathbb{R}$.

DEFINITION 2.3. *Let D_n be an open set of $\mathbb{R}^n$.*
$T^\star D_n = D_n \times (\mathbb{R}^n)^\star$ is a fiber bundle on D_n called **cotangent fiber bundle.**

DEFINITION 2.4. *Let $f : D_m \to D_n$ and let $F = D_n \times E$ be a fiber bundle, where E is a vector space. We define his* **pull-back**: $f^{-1}F = D_m \times E$

$$\begin{array}{ccc} f^{-1}F = D_m \times E & \stackrel{f\times Id}{\longrightarrow} & F = D_n \times E \\ (x,\ \mathbf{w}) & & (y,\ \mathbf{w}) \\ \downarrow & & \downarrow \\ D_m & \stackrel{f}{\longrightarrow} & D_n \\ x & \longmapsto & y = f(x) \end{array}$$

Let us apply this last definition to the cotangent fiber bundle previously introduced: $T^\star D_n = D_n \times (\mathbb{R}^n)^\star$. If $f : D_m \longrightarrow D_n$ is a differentiable mapping then we deduce:

$$\begin{array}{ccc} f^{-1}(T^\star D_n) & \longrightarrow & T^\star D_n \\ \downarrow & & \downarrow \\ D_m & \stackrel{f}{\longrightarrow} & D_n \end{array}$$

DEFINITION 2.5. *We associate to f a morphism of cotangent fiber bundle, denoted $T^\star f$, as following:*

$$\begin{array}{ccccc} T^\star D_m & \stackrel{T^\star f}{\longleftarrow} & f^{-1}(T^\star D_n) & \longrightarrow & T^\star D_n \\ (x,\ \mathbf{v}^\star) & \longleftarrow & (x,\ \mathbf{w}^\star) & & \\ \searrow & & \swarrow & & \downarrow \\ & D_m & & \stackrel{f}{\longrightarrow} & D_n \end{array}$$

with $\mathbf{w}^\star \in (\mathbb{R}^n)^\star$, $\mathbf{v}^\star \in (\mathbb{R}^m)^\star$ and $\mathbf{v}^\star$ defined using the transposed jacobian matrix of f with respect to x. Then $\mathbf{v}^\star = {}^t(T_x f).\mathbf{w}^\star$.
$T^\star f$ is called the **cotangent linear mapping**.

The composition of two cotangent linear mappings can be performed only when all the fiber bundles are over the same open set. More precisely, let $f = f_2 \circ f_1$ be such that

$$D_{n_1} \stackrel{f_1}{\longrightarrow} D_{n_2} \stackrel{f_2}{\longrightarrow} D_{n_3}$$

We consider $f_1{}^{-1}(T^\star D_{n_2})$, $f^{-1}(T^\star D_{n_3})$ and $T^\star D_{n_1}$. Thus we obtain three fiber bundles over D_{n_1}. Then we associate to them two mappings:

$$\begin{array}{ccccccc} T^\star D_{n_1} & \longleftarrow & f_1^{-1}(T^\star D_2) & \longleftarrow & f^{-1}(T^\star D_{n_3}) & & \\ \searrow & & \downarrow & \swarrow & & & \\ & & D_{n_1} & \stackrel{f_1}{\longrightarrow} & D_{n_2} & \stackrel{f_2}{\longrightarrow} & D_{n_3} \\ & & x_1 & \longmapsto & x_2 & \longmapsto & x_3 \end{array}$$

Developping this diagram we get:

$$\begin{array}{ccccccc} D_{n_1} \times (\mathbb{R}^{n_1})^\star & \stackrel{T^\star f_1}{\longleftarrow} & D_{n_1} \times (\mathbb{R}^{n_2})^\star & \stackrel{f_1^{-1}(T^\star f_2)}{\longleftarrow} & D_{n_1} \times (\mathbb{R}^{n_3})^\star & & \\ & & (x_1,\ {}^tT_{x_2}f_2.v_3^\star) & \longleftarrow & (x_1,\ v_3^\star) & & \\ & & \| & & & & \\ (x_1,\ {}^tT_{x_1}f_1.v_2^\star) & \longleftarrow & (x_1,\ v_2^\star) & & & & \\ & \searrow & \downarrow & \swarrow & & & \\ & & D_{n_1} & \stackrel{f_1}{\longrightarrow} & D_{n_2} & \stackrel{f_2}{\longrightarrow} & D_{n_3} \\ & & x_1 & \longmapsto & x_2 & \longmapsto & x_3 \end{array}$$

LEMMA 2.2. *Cotangent linear mapping of a composed function:*
if $f = f_2 \circ f_1$ then $T^\star f = T^\star f_1 \circ f_1^{-1}(T^\star f_2)$

We encounter two main difficulties when computing the cotangent linear mapping of a composed function: we have to keep track of the whole variables trajectory and to compose the cotangent linear mappings in reverse order with respect to the initial functions.

This mathematical formalism shows that automatic differentiation transformations on a program can be performed by composing elementary operations on the initial program statements. We now present its application to different kind statements for a language like FORTRAN.

2.2. Application to a sequence of assignments

EXAMPLE 2.1. *Let us consider the following sequence of assignments:*

$$\begin{aligned} z_1 &:= y_1 * y_2 \\ z_2 &:= z_1 + y_3 * y_4 \\ res &:= z_1/z_2 \end{aligned}$$

We assume that the inputs are y_1, y_2, y_3 and y_4 and that the output is res. This sequence of assignments computes a function $f : D_4 \to \mathbb{R}$.

In the functional representation previously described we get:

$$\begin{aligned} \tilde{f}_1 : D_{2+0+2} = D_4 &\to D_{2+0+1} = D_3 \\ (y_3, y_4, y_1, y_2) &\to (y_3, y_4, z_1) \\ & \quad z_1 = y_1 * y_2 \end{aligned}$$

The notation D_{2+0+2} represents that the variables y_3 and y_4 are not used by the assignment f_1 and are kept in memory for later use (they appear later in the program), there are no variables used and kept, and y_1 et y_2 are used and will not be useful later in the program then are not conserved.

We define by the same way $\tilde{f}_2$ and $\tilde{f}_3$.

$$\begin{aligned} \tilde{f}_2 : D_{0+1+2} = D_3 &\to D_{0+1+1} = D_2 \\ (z_1, y_3, y_4) &\to (z_1, z_2) \\ & \quad z_2 = z_1 + y_3 * y_4 \\ \tilde{f}_3 : D_{0+0+2} = D_2 &\to D_1 \\ (z_1, z_2) &\to res \\ & \quad res = z_1/z_2 \end{aligned}$$

2.2.1. Tangent linear code

The tangent linear mapping of $\tilde{f}$ which represents an assignment of the initial code can be computed as following:

$$\begin{aligned} T\tilde{f} : D_{n+p+s} \times \mathbb{R}^{n+p+s} &\longrightarrow D_{n+p+1} \times \mathbb{R}^{n+p+1} \\ (x,\ y,\ z,\ \mathbf{v}) &\longmapsto (x,\ y,\ t,\ \mathbf{w} = T_{(x,y,z)}\tilde{f}.\mathbf{v}) \end{aligned}$$

where f is given by a formula depending on the variables y and z.

Let F be a function computed by a sequence of assignments. F is a composition of functions $\tilde{f}_i$, each of them representing an assignment. $F = \tilde{f}_N \circ \tilde{f}_{N-1} \circ \ldots \circ \tilde{f}_1$

PROPOSITION 2.1. *The tangent linear code is produced by taking each assignment from the initial code and putting it in the tangent linear code predeceased by its corresponding "tangent linear assignment".*

Let us go back to the example 2.1 and compute tangent linear mappings associated to functions $\tilde{f}_1$, $\tilde{f}_2$ and $\tilde{f}_3$.

$$
\begin{aligned}
T\tilde{f}_1 : D_4 \times \mathbb{R}^4 &\rightarrow D_3 \times \mathbb{R}^3 \\
((y_3, y_4, y_1, y_2), (v_3, v_4, v_1, v_2)) &\rightarrow ((y_3, y_4, z_1), (v_3, v_4, w)) \\
& \quad w = y_2 v_1 + y_1 v_2
\end{aligned}
$$

which can also be written:

$$
\begin{pmatrix} v_3 \\ v_4 \\ w \end{pmatrix} = \begin{pmatrix} 1 & 0 & 0 & 0 \\ 0 & 1 & 0 & 0 \\ 0 & 0 & y_2 & y_1 \end{pmatrix} \begin{pmatrix} v_3 \\ v_4 \\ v_1 \\ v_2 \end{pmatrix}
$$

In the tangent linear code we just put the assignment computing the new value of w, the $v_i := v_i$ are useless from a computer point of view.

Then we compute $T\tilde{f}_2$ and $T\tilde{f}_3$:

$$
\begin{aligned}
T\tilde{f}_2 : D_3 \times \mathbb{R}^3 &\rightarrow D_2 \times \mathbb{R}^2 \\
((z_1, y_3, y_4), (w, v_3, v_4)) &\rightarrow ((z_1, z_2), (w, u)) \\
& \quad u = w + y_4 v_3 + y_3 v_4
\end{aligned}
$$

$$
\begin{pmatrix} w \\ u \end{pmatrix} = \begin{pmatrix} 1 & 0 & 0 \\ 1 & y_4 & y_3 \end{pmatrix} \begin{pmatrix} w \\ v_3 \\ v_4 \end{pmatrix}
$$

$$
\begin{aligned}
T\tilde{f}_3 : D_2 \times \mathbb{R}^2 &\rightarrow D_1 \times \mathbb{R} \\
((z_1, z_2), (w, u)) &\rightarrow (res, \eta) \\
& \quad \eta = \frac{1}{z_2} w - \frac{z_1}{z_2^2} u
\end{aligned}
$$

$$
\eta = \begin{pmatrix} \frac{1}{z_2} & -\frac{z_1}{z_2^2} \end{pmatrix} \begin{pmatrix} w \\ u \end{pmatrix}
$$

And the corresponding tangent linear code:

- Inputs: y_1, y_2, y_3, y_4, v_1, v_2, v_3, v_4
- Intermediate variables: z_1, z_2, w, u
- Outputs η, res

$$
\begin{aligned}
w &= y_2 v_1 + y_1 v_2 \\
z_1 &= y_1 * y_2 \\
u &= w + y_4 v_3 + y_3 v_4 \\
z_2 &= z_1 + y_3 * y_4 \\
\eta &= \frac{1}{z_2} w - \frac{z_1}{z_2^2} u \\
res &= \frac{z_1}{z_2}
\end{aligned}
$$

2.3. Adjoint code

Cotangent linear mappings of the function $\tilde{f}$ representing an assignment from the initial code are computed as following:

$$
\begin{array}{ccccc}
D_{n+p+1} \times (\mathbb{R}^{n+p+1})^\star & \longleftarrow & D_{n+p+s} \times (\mathbb{R}^{n+p+1})^\star & \xrightarrow{T^\star \tilde{f}} & D_{n+p+s} \times (\mathbb{R}^{n+p+s})^\star \\
 & & (x,\ y,\ z,\ \mathbf{w}^\star) & \longmapsto & (x,\ y,\ z,\ \mathbf{v}^\star = {}^t T_{(x,y,z)} f . \mathbf{w}^\star) \\
 & & \searrow & & \swarrow \\
(x,\ y,\ t) & \longleftarrow & & (x,\ y,\ z) &
\end{array}
$$

The matrix ${}^tT_{(x,y,z)}f$ is obtained by transposing the matrix $T_{(x,y,z)}f$. One assignment in the initial code produces $p+s$ assignments in the adjoint code. (The other n assignments $v_i^\star := w_i^\star$ are useless up to renaming).

Let F be a function computed by a sequence of assignments. F is a composition of functions $\tilde{f}_i$, each of them representing an assignment.

$$F = \tilde{f}_N \circ \tilde{f}_{N-1} \circ \ldots \circ \tilde{f}_1$$

By applying lemma 2.2, we find that the cotangent linear application of F is equal to

$$T^\star F = T^\star \tilde{f}_1 \circ f_1^{-1}(T^\star \tilde{f}_2) \circ \ \ldots \ \circ (f_{N-1} \circ \ldots \circ f_2 \circ f_1)^{-1}(T^\star \tilde{f}_N)$$

Let us consider again the sequence of assignments of example 2.1, functions $\tilde{f}_i$,and their associated tangent linear mappings $T\tilde{f}_i$. We compute the corresponding cotangent linear mappings. According to lemma 2.2 the composition of cotangent linear mappings must be done in the reverse order with respect to the initial functions. We aim to describe the mapping:

$$\begin{aligned} T^\star f:\ & D_4 \times I\!R^\star \rightarrow D_4 \times (I\!R^4)^\star \\ & (y_i, \eta^\star) \mapsto (y_i, v_i''^\star) \end{aligned}$$

The pull-back of $T\tilde{f}_3$ gives:

$$\begin{aligned} (f_2 \circ f_1)^{-1}(T^\star \tilde{f}_3):\ & D_4 \times I\!R^\star \rightarrow D_4 \times (R^2)^\star \\ ((y_3, y_4, y_1, y_2),\ \eta^\star) & \rightarrow ((y_3, y_4, y_1, y_2),\ (w^\star, u^\star)) \\ & {}^tT_{(z_1,z_2)}f_3.\eta^\star = (w^\star, u^\star) \end{aligned}$$

where $(z_1, z_2) = (f_2 \circ f_1)(y_3, y_4, y_1, y_2)$.

Writing explicitly the transpose matrix of the jacobian of f_3 with respect to z_1 and z_2, we get:

$$\begin{pmatrix} w^\star \\ u^\star \end{pmatrix} = \begin{pmatrix} \frac{1}{z_2} \\ -\frac{z_1}{z_2^2} \end{pmatrix} \eta^\star$$

We can observe that the whole trajectory has been computed to get the values of z_1 and z_2, and that the input variables are propagated during all of the process.

Similarly we compute:

$$\begin{aligned} f_1^{-1}(T^\star \tilde{f}_2):\ & D_4 \times (I\!R^2)^\star \rightarrow D_4 \times (I\!R^3)^\star \\ ((y_3, y_4, y_1, y_2),\ (w^\star, u^\star)) & \rightarrow ((y_3, y_4, y_1, y_2),\ (w'^\star, v_3'^\star, v_4'^\star)) \\ & {}^tT_{(z_1,y_3,y_4)}f_2.(w^\star, u^\star) = (w'^\star, v_3'^\star, v_4'^\star) \end{aligned}$$

$$\begin{pmatrix} w'^\star \\ v_3'^\star \\ v_4'^\star \end{pmatrix} = \begin{pmatrix} 1 & 1 \\ 0 & y_4 \\ 0 & y_3 \end{pmatrix} \begin{pmatrix} w^\star \\ u^\star \end{pmatrix}$$

$$\begin{aligned} T^\star \tilde{f}_1:\ & D_4 \times (I\!R^3)^\star \rightarrow D_4 \times (I\!R^4)^\star \\ ((y_3, y_4, y_1, y_2),\ (v_3'^\star, v_4'^\star, w'^\star)) & \rightarrow ((y_3, y_4, y_1, y_2),\ (v_3''^\star, v_4''^\star, v_1''^\star, v_2''^\star)) \\ & {}^tT_{(y_3,y_4,y_1,y_2))}f_1.(v_3'^\star, v_4'^\star, w'^\star) = (v_3''^\star, v_4''^\star, v_1''^\star, v_2''^\star) \end{aligned}$$

$$\begin{pmatrix} v_3''^\star \\ v_4''^\star \\ v_1''^\star \\ v_2''^\star \end{pmatrix} = \begin{pmatrix} 1 & 0 & 0 \\ 0 & 1 & 0 \\ 0 & 0 & y_2 \\ 0 & 0 & y_1 \end{pmatrix} \begin{pmatrix} v_3'^\star \\ v_4'^\star \\ w'^\star \end{pmatrix}$$

The last step consists in composing these morphisms.

Corresponding adjoint code:

- Inputs: y_1, y_2, y_3, y_4, $\eta^\star$
- Intermediate variables: z_1, z_2, $w^\star$, $u^\star$, $w'^\star$, $u'^\star$
- Outputs: $v_1''^\star$, $v_2''^\star$, $v_3''^\star$, $v_4''^\star$

$$
\begin{aligned}
z_1 &= y_1 * y_2 \\
z_2 &= z_1 + y_3 * y_4 \\
w^\star &= \frac{1}{z_2}\eta^\star \\
u^\star &= -\frac{z_1}{z_2^2}\eta^\star
\end{aligned}
$$

$$
\begin{aligned}
z_1 &= y_1 * y_2 \\
w'^\star &= w^\star + u^\star \\
v_3'^\star &= y_4 u^\star \\
v_4'^\star &= y_3 u^\star
\end{aligned}
$$

$$
\begin{aligned}
v_3''^\star &= v_3'^\star \\
v_4''^\star &= v_4'^\star \\
v_1''^\star &= y_2 w'^\star \\
v_2''^\star &= y_1 w'^\star
\end{aligned}
$$

The $v_i''^\star$ are the values of the partial derivatives of f with respect to y_i.

Remark: The recomputation of z_1 value is useless since z_1 appeared linearly in the initial program. Moreover its value has not been modified during the process then does not need to be recomputed. When implementing these mathematical rules, we took into account such observations.

2.4. Extension to other statements

This formalism may be extended to other statements. For now, we consider a programming language composed of conditional statements, subroutine calls and loops executing a number of iterations not depending on stopping criteria.

2.4.1. Subroutine calls

The execution of a subroutine call produces the same result as if the block of instructions it is composed of would have been put directly in the initial program. Then we suppose that we know how to produce automatically the corresponding tangent linear code or the adjoint code of a given subroutine. We just have to examine how they will be used.

1. Tangent linear code
   G(x_1, x_2, y) → TG(x_1, v_1, x_2, v_2, y, w)
2. Adjoint code
   G(x_1, x_2, y) → Gad (x_1, x_2, $w^\star$, $v_1^\star$, $v_2^\star$)

   with the same trajectory values as in the initial code.

We can remark that in some cases the called subroutine cannot be "differentiated" by applying transformation to its code (for example, if it comes from a library). In such cases, the appropriate subroutines should be called in the tangent linear code and in the adjoint code.

2.4.2. Conditional statements

We consider that conditional statements represent *piecewise differentiable functions* where each interval is an open subset of $\mathbb{R}^n$ defined by a conjonction of a finite number of constraints. A conditional statement is composed of three parts: a test and two branches. The test represents the interval on which the function is defined and each branch computes a composition of function on each interval.

We then deduce the rules to apply to a conditional statement to obtain the corresponding

1. tangent linear code

```
If test                If test
then B1      →         then (B1 tangent linear code)
else B2                else (B2 tangent linear code)
```

2. adjoint code

```
If test                If test
then B1      →         then (B1 adjoint code)
else B2                else (B2 adjoint code)
```

In both cases the test is evaluated on the same values in the initial code and in the statements computing the derivatives.

2.4.3. Loops

For now, in order to stay in the class of programs representing piecewise differentiable functions, we consider loops without stopping criteria depending on the execution of the program.

It can be considered that a loop is just an easy way to write a sequence of statements. Thus it can be extended, treated and the result can again be written using a loop.

1. Tangent linear code

   By applying the rules previously described on a sequence of statements we deduce how to produce the tangent linear code of a function represented by a loop:

```
Do i = 1, k              Do i = 1, k
   F code       →           TF code
   End Do                   End Do
```

2. Adjoint code

   As previously explained, we need to know all the trajectory. If we suppose that we compute it again, without saving intermediate computations, we have to create an internal loop to compute it. It is also necessary to invert the bounds of the loop (it means the way in which it is traversed) as the functions are composed in the reverse order with respect to the initial code.

PROPOSITION 2.2. *the adjoint code of a loop:*

```
Do i = 1, k
   F code
End Do
becomes
Do i = k, 1
   Do j = 1, i − 1
      F code
   End Do
   ᵗTₓF code
End Do
```

where x_{i-1} is the result of the internal loop (increment variable j) and the code of ${}^tT_{x_{i-1}}F$ is obtained by recursively applying the previous results about the adjoint code computation for each kind of statements.

Proof

- The initial loop (whose the adjoint code is computed) performs at each iteration:

  $$x_i = F_i \circ \ldots \circ F_1(x)$$

  where x represents the input value of the program. After the last iteration, we get: $x_k = G(x)$.
- To compute

  $${}^tT_xG = {}^tT_xF_1 \circ {}^tT_{F_1(x)}F_2 \circ \ldots \circ {}^tT_{(F_{k-1}\circ \ \circ F_1)(x)}F_k$$

  we compute at the ith step of the external loop:

  $$w_i^\star = {}^tT_{(F_{i-1}\circ .. \circ F_1)(x)}F_i(w_{i+1}^\star)$$
- Thus we compute with an internal loop: $x_{i-1} = F_{i-1} \circ \ldots \circ F_1(x)$

- Then w_i^* with respect to w_{i+1}^*.

Particular case: if there exist functions H_j such that
$\forall i, \ F_{i-1} \circ \ldots \circ F_1(x) = H_i \circ \ldots \circ H_n(y)$
then the corresponding adjoint code is:
Do i = 1, k
  $x := F(x)$
End Do
Do i = k, 1
  $x := H(x)$
  ${}^tT_xF$ code
End Do
For example, if F is a bijective function and H its inverse function, we have:

$$F_{i-1} \circ \ldots \circ F_1(x) = H_i \circ \ldots \circ H_n(y)$$

More particular case: if the values computed at each iterations do not depend on values computed at previous or next steps, trajectory variables may be computed in a loop performing iterations in the reverse order. Then trajectory variables and transposed linear forms are computed in the same loop.

3. The Odyssée system

3.1. Presentation

The Odyssée system is a complete toolkit of parametrised components [Rostaing and Dalmas, 1992] for FORTRAN 77 programs manipulation and differentiation. It has been written in a high level programming language, Caml-Light. It takes as input a FORTRAN program and parse it to represent each of its units (subroutine, function or main program) as a Caml object composed of an abstract syntax tree and a symbol table which contains information about all the occurring identifiers and their type. As we will describe it later, others tables are also used to store useful global datas. All the analysis and transformations are then performed on those structures, and the system returns as output a new FORTRAN program.

3.2. Differentiation of a unit

The main tool provided by the Odyssée system allows the user to manipulate a single subroutine which computes a function to get a new subroutine computing its derivatives. For this purpose, the user has to provide the list of all the active variables i.e. the input of the subroutine with respect to which we want to differentiate the function. The local dependence analysis then automatically determines for each statement the list of all the active variables which appears in this statement, that is to say the variables whose value depends on the input active parameters in at least one execution path when this statement will be executed.

3.2.1. The control graph

To perform this analysis, the program first computes the control graph of the subroutine, i.e. a description of all the possible execution paths. because, for now, no semantic can been associated to a program using GOTO statement, this graph only contains straight-line blocks, alternative blocks and loops.

Let us consider the simple example presented in Figure 1 which computes $(\frac{P(x)}{Q(x)})^2$ with

$$P(x) = \sum_{i=0}^{m} a_i x^i$$

$$Q(x) = \sum_{i=0}^{n} b_i x^i$$

The Figure 2 is the corresponding control graph.

```
Subroutine f ( a , m , b , n , x , y )
Dimension a(*)
Dimension b(*)
  r = 0.                          [1]
  s = 0.                          [2]
  t = 1.                          [3]
  Do i = 0, m                     [4]
    r = r + a(i) * t                 [4.1]
    t = t * x                        [4.2]
  EndDo
  t = 1.                          [5]
  Do i = 0, n                     [6]
    s = s + b(i) * t                 [6.1]
    t = t * x                        [6.2]
  EndDo
  If ( s .EQ. 0. ) Then           [7]
    y = 0.                           [7.1.1]
  Else
    y = r / s                        [7.2.1]
  Endif
  y = y * y                       [8]
  Return                          [9]
End
```

Fig. 1. An example of a hierarchic labelled fortran program

3.2.2. The dependence analysis

Using this graph, the Odyssée system then adds to each statement s with the collection a_s of all the "potentially active" variables when this statement will be executed. We call "potentially active" variable an identifier which value can depend on the input active parameters in at least one execution path (and so with respect to with the statement will be differentiated) .

Obviously, the status of an identifier can change only when the value of this identifier changes, so only if this identifier is the left side of an assignment or a parameter in a CALL statement. If s is a CALL statement, we must determine for each parameter if either its value is not changed, either its value is changed but depends on only unactive variables, or its value is changed and depends on at least one active variable. Therefore, the user has to describe this information in a table containing for each unit :

- its name
- a mask $(mask_i)_{i=1...n}$ describing which of its n parameters have to be active
- its differentiating method

This table provides to Odyssée all the needed information to perform the dependence analysis on a unit using lower-level subroutines whose source codes are not available.

If we define u_s as the union of the active variables sets of all the parents of s in the control graph, the updated list a_s needed for the dependance analysis can then be defined by the following rules :

- if s is an assignment, we consider
  - var = the left part of s
  - $list$ = all the variables used in the right part of s

  then we have :
  1. if ($(u_s \cap list = \emptyset)$ & (var is not an array))
     then $a_s = u_s - \{var\}$
  2. if $(u_s \cap list \neq \emptyset)$
     then $a_s = u_s \cup \{var\}$

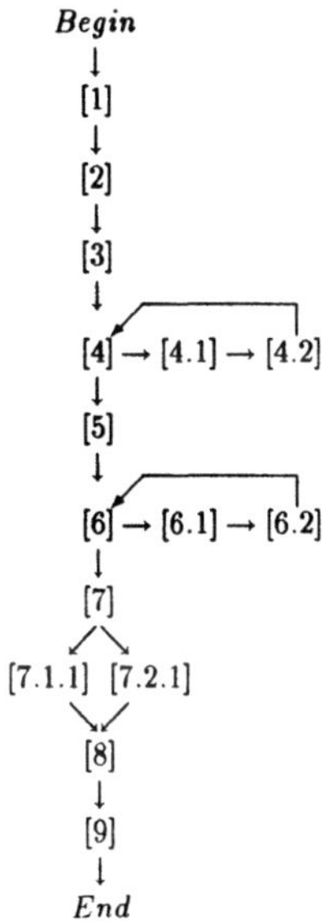

Fig 2 An example of control graph

the array test in (1) prevents an active array to become inactive just because one of its items becomes inactive.

- if s is a CALL statement, we first set $a_s = u_s$ and then we compare each parameter's status with the information given in the differentiation method table
  1. if (($param_i$ is in u_s) & ($mask_i$ is unactive))
     then an error is raised : the differentiation method table has to be adapted.
  2. if (($param_i$ is not inu_s) & ($mask_i$ is active))
     then $a_s = a_s \cup \{param_i\}$
  3. else a_s is not changed
- else, $a_s = u_s$.

The Figure 3 describes the result of such an analysis when we expect to calculate the derivative of f (see Figure 1) with respect to x.

As seen in the former section, each statement can then be differentiated, according to the rules described either in the tangent linear code method if we wish a program which computes the directional derivatives, or in the adjoint code method to get a program which computes all the partial derivatives.

We can remark that some initialisation statements have to be added each time several execution paths join together but have not the same active variables list. For example, in Figure 2, the y variable is active in one branch of the If but inactive in the other one. y is so considered as a active variable in the [8] statement. When differentiating this statement, the dy variable will be used whereas it has not been initialized if the execution path has been :

$$\ldots \rightarrow d[7.1.1] \rightarrow d[8] \rightarrow \ldots$$

As we can see in Figure 4, the d[7.1.1] statement had to be completed with $dy = 0.$ as if y was an active variable in this statement.

```
Subroutine f ( a , m , b , n , x , y )
Dimension a(*)
Dimension b(*)
  r = 0.                             {x}
  s = 0.                             {x}
  t = 1.                             {x}
  Do i = 0, m                        {x;r;t}
    r = r + a(i) * t                   {x;r;t}
    t = t * x                          {x;r;t}
  EndDo
  t = 1.                             {x;r}
  Do i = 0, n                        {x;r;s;t}
    s = s + b(i) * t                   {x;r;s;t}
    t = t * x                          {x;r;s;t}
  EndDo
  If ( s .EQ. 0. ) Then              {x;r;s;t}
    y = 0.                             {x;r;s;t}
  Else
    y = r / s                          {x;r;s;t;y}
  Endif
  y= y * y                           {x;r;s;t;y}
  Return                             {x;r;s;t;y}
End
```

Fig. 3. An example of dependence analysis

All the new identifiers are added in the declaration part and the subroutine parameters are completed with both the input and output active variables to obtain a new compilable unit in output. Figure 4 presents result of the direct mode method in computing derivatives of the Figure 1 example.

The Odyssée structure allows the user to parametrise the differentiation with its own strategy, that is to say to a choice between what must be saved and what must be recomputed. We present in the next section one possible strategy already implemented for the ARPEGE/IFS code.

Application to the ARPEGE/IFS *code*

The ARPEGE/IFS program has been written with the objective to use an automatic differentiation tool to produce the tangent linear and the adjoint codes. Then it respects some programming standards and some FORTRAN restrictions. The strategy (compromize between values stored and recomputed when needed in the derivatives) is adapted to this code and is already planned in the organisation of the program representing the numerical model [Rostaing, Dalmas and Galligo, 1993].

The tangent linear code used in the ARPEGE/IFS program computes directional derivatives and the whole initial function is not needed at this stage. As the initial function and the tangent linear application are computed in two different compilation units, it is necessary to give to the subroutine computing the derivatives the part of the trajectory needed in the derivatives, i.e. the values involved in non linear computations. The choosen strategy consists in computing again the values locally computed in the initial program and giving as parameters the values which were given as parameters in the initial program.

The strategy is the same in the adjoint code. But in this case parts of the trajectory may be computed several times or stored since the transpose linear forms are computed in the reverse order and the trajectory in the same order with respect to the initial code. Let us consider again proposition 2.2. Instead of computing several times parts of the trajectory into an internal loop, the trajectory is computed and saved in a local array. Each time the more particular case can be applied, the needed trajectory values are computed in the same loop as the transposed linear forms.

This strategy has consequences on the knowledge basis. The adaptation of this basis to the ARPEGE/IFS

```
Subroutine df ( a , m , b , n , x , y , dx , dy )
Dimension a(*)
Dimension b(*)
  r = 0.
  s = 0.
  t = 1.
  dr = 0.
  ds = 0.
  dt = 0.
  Do i = 0, m
    dr = dr + a(i) * dt
    r = r + a(i) * t
    dt = dt * x+t * dx
    t = t * x
  End Do
  t = 1.
  dt = 0.
  ds = 0.
  Do i = 0, n
    ds = ds + b(i) * dt
    s = s + b(i) * t
    dt = dt * x+t * dx
    t = t * x
  EndDo
  If ( s .EQ. 0. ) Then
    y = 0.
    dy = 0.
  Else
    dy = ( ( dr * s ) - ( r * ds ) ) / ( s *s )
    y = r / s
  Endif
  dy = 2 * dy * y
  y = y * y
  Return
End
```

Fig. 4. An example of direct mode differentiation

code concerns the "differentiation method". In the ARPEGE/IFS code, we can apply three different methods to produce the derivative statement corresponding to a subroutine call:

- if the subroutine called computes a linear function, the same subroutine is called in the "derivatives" programs.
- if the "derivative" subroutine needs values computed by the initial subroutine in its list of parameters, a subroutine call in the initial code gives two calls in the tangent linear code or the adjoint code:
  - the same statement as in the initial subroutine in order to compute trajectory values needed by the derivatives.
  - the corresponding tangent linear or adjoint subroutine is called with the corresponding list of arguments

  This rule can be applied when the output values are modified only once. It is an example of the adaptation of the initial numerical model to the choosen strategy.
- else, the corresponding tangent linear or adjoint subroutine is called with the corresponding list of arguments. This list is built according to an information also given by the user: which variables (in

the mathematical meaning) will not be involved in non linear computations. They will not be given as parameters to the derivative subroutines.

The ARPEGE/IFS strategy has been completely implemented in *Odyssée* ([Rostaing, Dalmas and Galligo, 1993]) The produced code is similar to the hand-written code whose rules are described in [Talagrand, 1991].

3.3. Differentiation of a collection of units

Models we wish to compute derivatives are most of the time represented by large FORTRAN programs composed of a main subroutine using a set of lower-level subroutines. With the object of facilitating the manipulation and differentiation of those more complex structures, new tools has been developed in Odyssée. Their main purpose is to automatically fill the required differentiation method database, usually completed by the user, when this information can be deduced by our system that is to say when all the source code is available. The user has just to indicate the input active variables of the main subroutine, and the system automatically determines all the active variables for all the lower-level subroutines.

3.3.1. The flow graph

To be able to perform an interprocedural analysis, the Odyssée system first determine the collection organization. The list of all the parsed units is completed with a graph whose edges are those units and vertices translate the relation "calls". For a given unit, the system can then very easily recursively extract all the needed lower-level subroutines and functions. It detects all the required units whose source code is not available. Finally, it looks for direct or indirect recursive definitions (i.e. "f calls ... calls f") which are forbidden in FORTRAN 77. The obtained graph can then be printed as a tree thanks to a specific tool. We present in Figure 5 the result of this analysis on a sample of the ARPS model ([CAPS, 1992]), a $\simeq$ 150000 lines FORTRAN program for storm forecast. The parenthesised named are those whose source code has not been provided.

3.3.2. The interprocedural data dependence analysis

To perform this analysis, the user has to provide :

1. for the top unit, the source code and the input active variables
2. for the lower-level units, either the source code and the input active variables, or all the input and output active variables

The local analysis of the top unit is then performed as previously described, but with this adapted case if s is a CALL statement :

- we first set $a_s = u_s$ and then we compare each parameter status with the information given in the differentiation method table
    1. if (($param_i$ is in u_s) & ($mask_i$ is unactive)) then
        - new_mask_i is set to active
        - a_s is not changed
    2. if (($param_i$ is not in u_s) & ($mask_i$ is active)) then
        - $a_s = a_s \cup \{param_i\}$
        - $new\_mask_i = mask_i$
    3. else
        - $a_s = a_s$
        - $new\_mask_i = mask_i$
- - if $new\_mask = mask$ then the system keeps on the analysis
    - if $new\_mask \neq mask$ and the called unit source code is not available then an error is raised
    - if $new\_mask \neq mask$ and the called unit source code is available then
        * the called unit is analysed with its new mask

let arps = get_program
["solve3d.f"; "force3d.f"; "operat3d.f"; "step3d.f"; "bc3d.f"; "tinteg3d.f"; "divg3d.f"; "setbdt3d.f"; "tmix3d.f"; "advct3d.f"; "cmix3d.f"; "init3d.f"; "raydmp3d.f"; "bcdif3d.f"; "micro3d.f"; "arps31.f"; "out3d.f"; "sfcphy3d.f"; "dump3d.f"; "outlib3d.f"; "svidmpstub.f"; "arpslib3d.f"; "inibase3d.f"; "rst3d.f"; "energy3d.f"]
;;
arps : *FortranProgram* =< *Abstract* >

print_graph arps ;;

```
--arps31----+-chkstab---+-(dtadump)
            |           +-(gtdmpfn)
            |           +-fmtprt------(wrigar)
            |           +-(grafclose)
            |           +-maxmin----+-(fnversn)
            |                       +-(a3dmax)
            +-cordintg--+-tinteg----+-solvqv----+-stepq
            |           |           |           +-mixqv-----+-cmix4q----+-difzz
            |           |           |           |           |           +-bsdifxx
            |           |           |           |           |           +-bsdifzz
            |           |           |           |           |           +-difyy
            |           |           |           |           |           +-bsdifyy
            |           |           |           |           |           +-difxx
            |           |           |           |           |           +-aamult
            |           |           |           |           +-cmix2q----+-difzz
            |           |           |           |           |           +-difyy
            |           |           |           |           |           +-difxx
            |           |           |           |           |           +-aamult
            |           |           |           |           +-flzero
            |           |           |           |           +-tmixqv------divgsg
            |           |           |           +-latbdtq
            |           |           |           +-bcq---------bcsclr
            |           |           |           +-advq------+-uvwrho
            |           |           |                       +-advcts----+-avgx
            |           |           |                                   +-avgy
            |           |           |                                   +-avgz
            |           |           |                                   +-difx
            |           |           |                                   +-dify
            |           |           |                                   +-difz
            |           |           |                                   +-aamult
            |           |           +-solvq-----+-stepq
            |           |           |           +-latbdtq
            |           |           |           +-bcq---------bcsclr
            |           |           |           +-advq------+-uvwrho
            |           |           |           |           +-advcts----+-avgx
            |           |           |           |                       +-avgy
```

Fig. 5. An sample of the ARPS flow graph

* if it was the first occurence of this unit
  then this modification has no other effect on the current state
  of the analysis
  else we have to restart the complete analysis of the top unit with the new adapted
  differentiation method database

A finest incremental algorithm could be implemented, but would probably require more detailed information, and in particular the complete dependence graph, to prevent the system from repeating the complete analysis.

4. Conclusion and perspectives

We have implemented the ARPEGE/IFS strategy in the *Odyssée* system. As *Odyssée* is a parametrised toolkit, we plan to adapt *Odyssée* to other applications with other specifities than ARPEGE/IFS. The *Odyssée* computer algebra functionalities will be improved to allow optimizations when evaluating the produced code. This work is in progress.

For now, the implemented strategy depends only on the identifiers use (local, parameter, global, ...). We can extract other strategies from the presented functional representation. In proposition 2.2, instead of recomputing all the trajectory at each step or saving every step in an array, it could be interesting to implement storage/recomputation strategies like for example saving a value every n steps and recomputing the intermediate values where criteria for choosing n have to be studied [Griewank, 1992].

The ARPS program has been developped to use parallelism possibilities. We are working with F.X. Le Dimet's team in order to adapt ODYSSÉE to ARPS and particularly the effect of parallel algorithms in the automatic development of the adjoint code.

Another domain of work is the extension of the class of considered functions. The functional representation has to be improved to take into account iterative processes. More generally we are lead to study functions implicitly defined as solutions of algebraic equations. A collaboration began with X. Joulia's team (LEAP, INP-ENSIGC Toulouse) to apply *Odyssée* to chemical process analysis.

References

Bischof et al., 1992 C. Bischof, A. Carle, G. Corliss, et A. Griewank. ADIFOR: Automatic Differentiation in a Source Translator Environment. In Paul S. Wang, editor, *Proceedings of the International Symposium on Symbolic and Algebraic Computation*, 1992.

CAPS, 1992 *ARPS version 3.0 User's Guide*. CAPS, University of Oklahoma, NSF, 1992.

Courtier et al., 1991 P. Courtier, C. Freydier, J.F. Geleyn, F. Rabier, and M. Rochas. The ARPEGE project at Météo-France. In *ECMWF Seminar Proceedings, Numerical methods in atmospheric models, Vol 2*, September 1991.

Giering, 1993 R. Giering. *Adjoint Model Compiler, Version 2.15*. Max-Planck Institut fur Meteorologie, Hamburg, Germany, 1993.

Griewank, Juedes et Srivasan, 1990 A. Griewank, D. Juedes, et J. Srivasan. ADOL-C, a package for the automatic differentiation of algorithm written in C/C++. Technical report, Argonne National Laboratory, Illinois USA, 1990.

Griewank, 1992 A. Griewank. Achieving logarithmic growth of temporal and spatial complexity in reverse automatic differentiation. *Optimization methods and softwares*, 1:35–54, 1992.

Le Dimet et Talagrand, 1986 F.X. Le Dimet et O. Talagrand. Variational algorithms for analysis and assimilation of meteorological observations: theoretical aspects. *Tellus*, numero 38A:97–110, 1986.

Rostaing, Dalmas and Galligo, 1993 N. Rostaing, S. Dalmas, and A. Galligo. Automatic Differentiation in Odyssée. *Tellus*, numero 45A, 1993.

Rostaing and Dalmas, 1992 N. Rostaing et S. Dalmas. Automatic FORTRAN programs analysis and transformation using a typed functional language. In R. Glowinski, editor, *Proceedings of the 10th International Conference on Computing methods in applied sciences and engineering*, pages 527–535. Nova Science Publisher, 1992.

Talagrand, 1991 O. Talagrand. The use of Adjoint Equations in Numerical Modelling of the Atmospheric Circulation. In G.Corliss A.Griewank, editor, *Automatic Differentiation of Algorithms: Theory, Implementation and Application*, pages 169–180, 1991.

Translation Tools for High Performance Computing

Matthew T. O'Keefe, and Aaron Sawdey
Army High Performance Computing Research Center
and
Department of Electrical Engineering
University of Minnesota
Minneapolis
Minnesota 55455
USA

ABSTRACT.
Massively parallel processors (MPPs) hold the promise of extremely high performance that, if realized, could be used to study problems of unprecedented size and complexity. One of the primary stumbling blocks to this promise has been the lack of tools to automatically translate application codes to MPP form. In this paper we show how applications codes written in a subset of Fortran 77 called **Fortran-P** can be translated to achieve good performance on several massively parallel machines. This subset can express codes that are *self-similar*, where the algorithm applied to the global data domain is also applied to each subdomain. We have found many codes that match the Fortran-P programming style and have converted them using our tools. We believe a *self-similar* coding style will accomplish what a *vectorizable* style has accomplished for vector machines by allowing the construction of robust, user-friendly translation systems that increase programmer productivity and generate fast, efficient code for MPPs.[1]

1. Introduction

Distributed-memory massively parallel processors (MPPs) consisting of many hundreds or even thousands of processors have become available and offer peak performance in the tens to hundreds of Gigaflops range [20], [23]. Achieving a significant fraction of this performance would allow the study of many important physical problems (compressible, turbulent flows [16], meso-scale weather prediction [7], ocean circulation, etc.) that would otherwise be intractable. Unfortunately, recent published results [11] have suggested that these performance levels are not yet sustainable on complete applications codes. The goal of the work described here is to achieve very high performance on MPPs while retaining portability and ease of programming for real applications codes.

[1] This work was supported in part by Office of Naval Research Grant No. N00014-93-1-0426, by contract no. DAAL02-89-C-0038 between the Army Research Office and the University of Minnesota for the Army High Performance Computing Research Center, National Science Foundation Grant NSF CCR-9110261, and by a grant from the Center for Analysis and Prediction of Storms (CAPS).

F.-X. Le Dimet (ed.), High Performance Computing in the Geosciences, 223–244.

The approach we propose is to exploit certain characteristics of application codes that map well to MPPs. These massively parallel applications are written in a subset of Fortran 77 we call Fortran-P and automatically translated to a language suitable for MPP execution. There are many codes, including PPM [25] and ARPS [7], that naturally fit the definition of Fortran-P and can be directly translated. The Fortran-P language was derived from the style used by Woodward in coding his hydrodynamics algorithms [25] for a variety of high performance computers.[2]

We have refined this language and implemented tools that translate Fortran-P programs into massively parallel codes running on the CM-200, CM-5, and other parallel machines. These are complete codes used daily at our site to do large, state-of-the-art calculations.

This paper is organized as follows: we first describe the Fortran-P model in an informal way. We then describe the structure of the current translator in the context of translating a global ocean model, followed by its use in semi-automatically converting the ARPS meso-scale weather code to CM Fortran. The last section gives conclusions and directions for future work.

2. The Fortran-P Model

Many large applications codes developed in serial languages such as Fortran 77 have been resistant to good vectorizing and parallelizing compiler technology simply because the underlying algorithms are inherently serial. There is a widespread—and to some extent justified—view that programmers will have to reconsider their basic algorithms for the new massively parallel environment. The result of this viewpoint has been a focus on language design rather than the translation of codes expressed in serial languages to MPP form [10]. The latter problem, especially the selection of an efficient data layout for a distributed memory machine, has generally been considered too difficult to automate and has been left to the programmer ([9] is an exception).

We do not claim that all codes written in traditional languages like Fortran 77 can be automatically translated for efficient execution on an MPP. Rather, we argue that many programmers have, over the years, developed an understanding of how to express their parallel algorithms in "serial" programming languages. In fact, this is precisely what is done when writing in a "vectorizable" style for a vector machine. Over time compiler technology, in the form of program analysis and recognition of common programming idioms, has matured to the point that vectorization is almost entirely automated. We believe that a similar approach using a *self-similar* programming style can allow efficient automatic parallelization of serial code for MPPs. The language target for the automatic translator could be data-parallel with data layout primitives, (such as CM Fortran [22] or HPF [10]), a message-passing model, or message-passing combined with a native code compiler for a node.

We have developed such a tool for a subset of Fortran 77 we call the **Fortran-P**.

[2] Woodward first developed several of these ideas when implementing PPM on a Cray using solid-state disks.

[3] Fortran-P codes are *self-similar*, allowing them to fully exploit the parallelism in an MPP, even as these machines scale up to ever larger sizes. Compilers can also exploit this property to automatically generate efficient data layouts and communication code.

The Fortran-P translator allows the MPP environment to mirror the typical vector supercomputer work environment in which codes are written and maintained in Fortran 77 and then vectorized (and often parallelized) as an integral part of the compilation process. In the Fortran-P environment, the programmer develops and maintains his codes on a workstation or a vector supercomputer and translates the program to MPP form only as a final step before running a large calculation where the enormous memory and computing capacity of an MPP are required. Programmers can take advantage of the mature and stable development environments available on these machines, debugging codes on smaller problems until the program is ready for a large production run on the MPP. This approach takes advantage of the portability of Fortran 77 across a wide variety of machines, from PCs to vector supercomputers.

In the next two sections, we describe the Fortran-P programming model and the translation process.

3. The Fortran-P Programming Model

Self-similar programs are written so that the computations applied to the whole domain are precisely the same as those applied to each subdomain.[4]

Programs written in Fortran-P perform computations over a logically homogeneous grid of zones where (in two dimensions) the data domain appears as shown in Figure 1. Here **NX** is the problem size in the x dimension, **NY** is the problem size in the y dimension, and `nbdy` is the number of boundary (fake) zones. Boundary zones are employed to implement several different kinds of boundary conditions [25]. Note that the current translator also supports one- and three-dimensional grids (higher dimension grids are possible).

In general, a self-similar program has the following properties:

1. *A logically regular computational grid.*
   An irregular grid would not be self-similar as subdomains could have different grid topologies. In Fortran-P, each subdomain grid must have a topology identical to that of the global grid.
2. *Dependence distances [24] are small constants that are independent of the problem size, which implies local communication operations in the translated code.*
   If dependence distances were a function of the grid size then when partitioning is performed, non-local communication would be required.[5] In Fortran-P, updates

[3] In fact, Fortran-P currently includes several Fortran 90 array intrinsics.

[4] This is analogous to the self-similarity of many fluid flows: intuitively, self-similarity means that properties that are apparent globally are also found in small subregions of the flow; hence, if an observer could simultaneously examine both the complete domain and a small subdomain of the flow (without clues to the actual domain sizes) then he or she would be unable to tell them apart.

[5] There is an exception. For *periodic* boundary conditions the dependence distance is the equal to the extent of the particular dimension. Boundary zone handling is discussed in a following

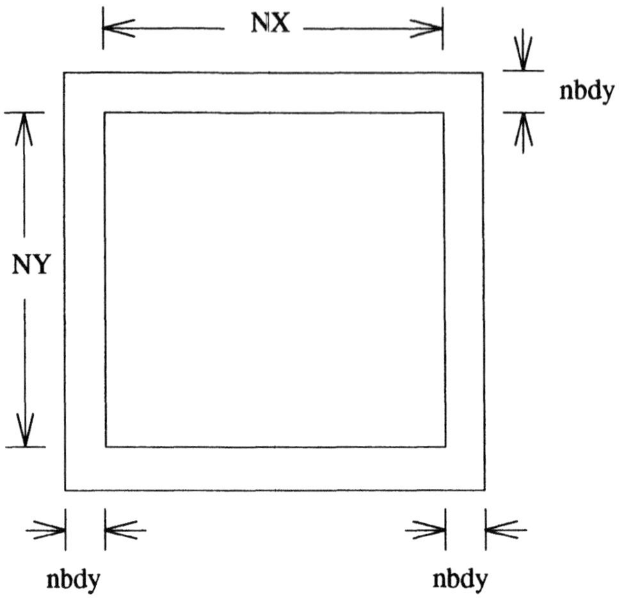

Fig. 1. A 2-D Data Domain with Surrounding Boundary Zones

of boundary zones imply that interior "fake" zones (to be described shortly) must be updated as well.

3. *Each subdomain requires nearly the same amount of computation.*
   The same algorithm is applied to equal-sized subdomains so, in general, the amount of computation will be nearly the same. Some variance will arise from data-dependent operations but this should be balanced out when subdomains are relatively large.
4. *Reduction operations are permitted, recurrences are not.*
   Reductions can be implemented in two steps: first, perform the reduction locally within each processor, and second, perform a global reduction step using the result from the local computation. Recurrences are not self-similar in the sense that each result depends on a different number of operations and the calculation is inherently sequential.

These properties permit the data domain of self-similar Fortran-P programs to be automatically partitioned and then distributed equally across all processors. Each processor is given the same amount of work to perform. Fortran-P programmers write code so that operations that apply locally also apply globally. The translator can then exploit parallelism at multiple levels, both across the machine and within a node.

Assuming a 3x3 processor grid the two-dimensional domain is partitioned as

section.

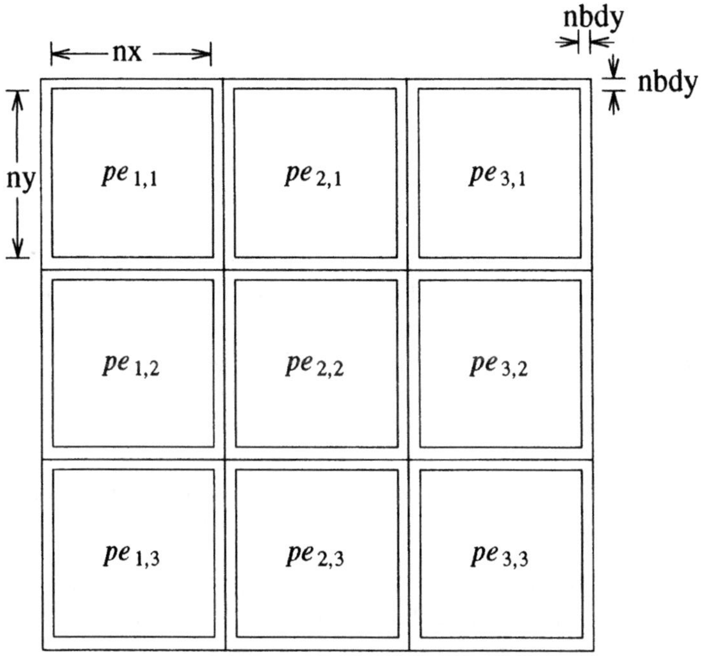

Fig. 2. Partitioning a 2-D Domain into 3x3 Patches

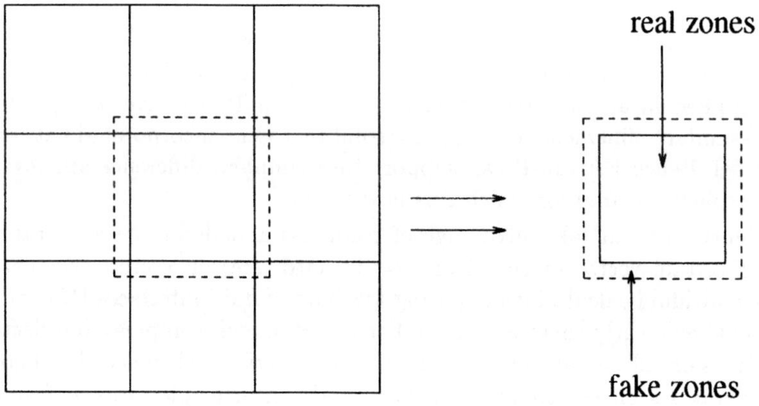

Fig. 3. Fake Zones for an Interior Subdomain

shown in Figure 2. Note how each interior subdomain (or *sub-patch*) is now surrounded by a picture frame of *fake zones*. These fake zones actually contain redundant data from neighboring processors, as shown in Figure 3.

Fake zones (also referred to as *overlap regions* [8]) allow computations to be performed using only local data; as the algorithm proceeds, the fake zones become corrupted and eventually must be updated using real zones from neighboring processors. In current Fortran-P codes, fake zone updates are done simultaneously with boundary zone computation. No interprocessor communication is generated for loops which range over real (*interior*) zones *as well as* fake zones. This allows the zones just inside the boundaries to be updated properly without any special boundary treatment, so long as the values in the fake zones are properly set.

Another advantage of fake zones is that they allow the compiler to trade off memory with communication; larger overlap regions reduce the frequency of communication but require more memory. Hence, the compiler could vary the overlap region size to obtain best performance on a given machine.[6]

In the current Fortran-P model boundary code is written so that fake zone updates are generated when a loop is found that writes *only* the boundary array elements. Fortran-P currently recognizes four kinds of boundary conditions: *reflecting, periodic, continuation*, and *prescribed value* [25]. More general boundary conditions could easily be supported within this Fortran-P framework.

Global operations such as finding a global minimum or maximum are sometimes required in Fortran-P programs; for example, in PPM, the Courant number must be calculated during each time step, and this requires a global maximum. The current translator calls fast, vendor-supplied intrinsics (when available) to perform global operations efficiently. These kind of global reductions can be performed most efficiently in a self-similar style: the reduction is performed locally on each processor, and the same operation is then applied globally across processors using the local results.

Certain kinds of elliptic equations requiring global information can be addressed with multigrid relaxation schemes, which are made up entirely of local, self-similar operations applied in an iterative sequence. At each multigrid level the problem remains self-similar; refinements to the grid resolution can be incorporated naturally into the model. Hence Fortran-P can support these implicit difference schemes as well as the explicit schemes for which it is now used.

Many algorithms can take advantage of compress and decompress operations (available on certain vector machines directly in hardware) on vectors. For example, certain multifluid calculations requiring tracking of fluid interfaces [12] can be computed most efficiently this way. In the Fortran-P model, compress and decompress operations on subdomain data can effectively be performed on each local node. Although operations on the compressed data might in principle proceed faster if this work load were rebalanced, no such dynamic load balancing is now envisioned for Fortran-P.

[6] For example, the CM-200 [19], which has four megabytes of memory per node and relatively slow network speed, would perform better with more fake zones than the Maspar MP-1 [1], which has only 64 Kbytes of memory per node but a relatively fast network.

4. The Fortran-P Translator

In this section we describe the current implementation of the Fortran-P translator in the context of translating the University of Miami Isopycnic Ocean Model. The Fortran-P translator is divided into phases, each phase constructed using a tool known as SORCERER [14]. This tool simplifies the construction of source-to-source translators by generating the actual translator from a grammar description of the tree transformations that represent the required translations.

The path that Fortran-P codes generally take through the translator is illustrated by the right side of Figure 4. The Miami Ocean Model reused some phases for a specialized translation (shown on the left of Figure 4) before being translated to CM Fortran. The names and functions of the various phases are listed below.

fort This phase translates Fortran 77 (Fortran-P) into the Intermediate Representation.

addpar This phase adds the additional "parallel" indices required by the domain decomposition data layout scheme to each array reference.

cdir This phase interprets directives and sets flags in IR nodes based on their information. Phases following this one can simply look at the flags; they do not have to find the comment directive that resides in a separate node from the one on which it acts.

fixlabels This phase moves `goto` and `do`-loop labels attached to statements to separate continue statements.

atlfix This custom phase is described in Section 4; it was written to modify the loop style used in the Miami model to enable MPP translation.

ifmerge Wherever possible, this phase converts `if` statements into Fortran 90 `MERGE` intrinsics; it is described in Section 4.

chains This phase builds symbol-reference chains. Each symbol has a chain, or list of pointers, pointing to all the nodes that reference it. This information is needed by the scalar promotion phase.

suds This phase builds simple def-use and use-def pointer information. Pointers are set from the definition of a variable to nodes where it is used, and from nodes where a variable is used back to the definition of the variable.

scalar This phase does scalar promotion. It requires the symbol-reference information created in the **chains** phase.

sbdy This phase translates boundary code delimited by special directives. It requires the use-def and def-use information created by the **suds** phase.

cmf This phase generates Fortran 77, Fortran 90, or CM Fortran from the IR.

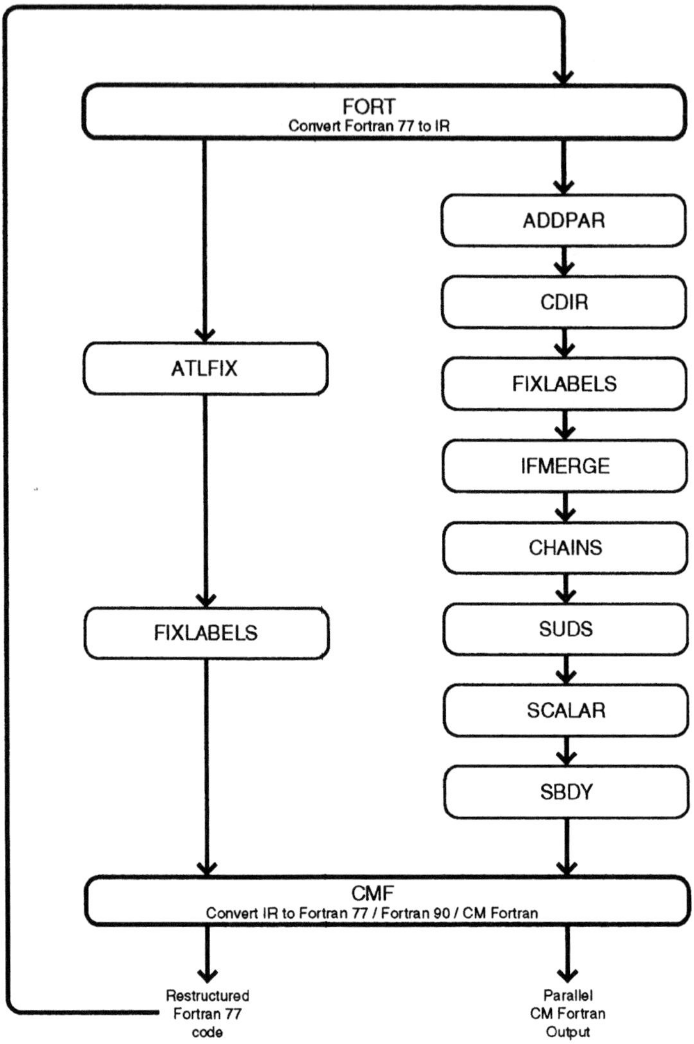

Fig. 4. The flow of a program through the translation system. The right side represents the general case of conversion from Fortran 77 code to CM Fortran code, while the left side shows how three of the phases were reused along with one specialized phase to do Fortran 77 to Fortran 77 code restructuring.

4.1. BUILDING THE TRANSLATOR WITH SORCERER

SORCERER generates C code with embedded user-specified actions that matches (or walks) trees. It provides an environment where programmers can avoid the tedious coding of tree parsers while not limiting themselves to describing the intended

```
ifcheck:  ( If expr
              ( SLIST ( Assign expr expr ) )
              ( SLIST ( Assign expr expr ) )
          )
```

Fig. 5. SORCERER grammar to match "MERGEable" `if-else` constructions.

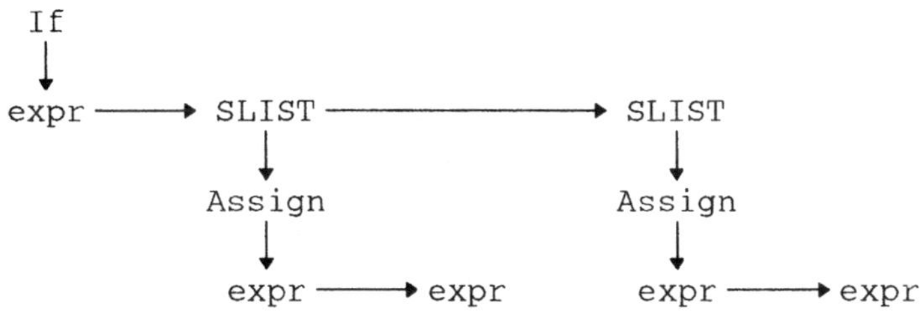

Fig. 6. Tree to match to find "MERGEable" `if-else` constructions

translation problem purely as tree transformations. These features are useful in three ways.

1. We have a "skeleton" SORCERER description that matches any valid Fortran 77 program as represented in the IR. This provides a solid foundation for writing new phases that annotate and/or modify programs represented in the IR since all that needs to be added are actions; all the tree-recognition is done automatically.
2. Since SORCERER generates C code, rules for matching special tree constructions can be included in the SORCERER description and called from actions in the SORCERER grammar or from other C functions. For example, we used this approach when matching an `if` statement with a single assignment statement in the `then` and `else` clauses. Such constructs can be converted to Fortran 90 `MERGE` statements (which are much faster on the CM-5 than `if-then-else`). In our SORCERER actions, we compare the left-hand-side of the assignment statements in the `then` and `else` clauses to determine if the `if` statement could be turned into a Fortran 90 `MERGE` intrinsic. The SORCERER description for this special match is shown in Figure 5, and the tree that this SORCERER grammar matches is shown in Figure 6.

```
expr:   Var << printf("%s",SYMBOL(this)); >>
      | ( "\+" expr << printf("+"); >> expr )
      | ( "\-" expr << printf("-"); >> expr )
      ;
```

Fig. 7. A SORCERER description for printing simple expressions.

3. SORCERER routines generate source code from the IR. In this case, the actions simply print out the destination-language equivalents for the tree-form constructs matched. For example, a SORCERER description for printing simple expressions is shown in Figure 7.

4.2. CM-FORTRAN

The current principal target MPP language for the translator is CM Fortran [22]. CM Fortran includes Fortran 90 array syntax and intrinsics as well as data layout directives. In CM-Fortran, scalar objects are resident on the front-end processor and array objects may reside on the front-end or be distributed across the processing nodes. Compiler directives are provided to allow the user to specify whether a dimension of a distributed array is within a node (`:SERIAL`) or spread across the nodes (`:NEWS`). This meshes well with the Fortran-P model; the original dimensions remain `:SERIAL` and the added dimensions are labelled as `:NEWS`. This way the serial dimensions address data elements within the subdomain and will ensure that the subdomains are not split between processors; the remaining dimensions form a processor address and specify the particular subdomain.

4.3. TRANSLATION IN HIGH PERFORMANCE COMPUTING ENVIRONMENTS

Current high performance computing environments are heterogeneous, often consisting of shared memory vector multiprocessors, high-end graphics workstations and servers, and distributed-memory massively parallel processors. At the Army High Performance Computing Research Center, we have access to an eight-processor Cray YMP-C90, a 16-processor SGI POWER Challenge machine, and a 544-node CM-5. We port applications codes between these machines and it is often necessary to exploit special features and capabilities or to avoid performance blind spots in their compilers; in addition, the codes themselves are often in different phases of development and change rapidly over the course of several weeks.

In the early phases of its development, a code might be written and maintained in Fortran 77 and run on Sun SPARC workstations. After the algorithms have been debugged, the same code might be compiled on multiprocessor SGI workstations to run larger problems. On the SGI workstations the code is still in Fortran 77, but additional directives can instruct the compiler how to parallelize the code and perform data layout. The Cray YMP-C90 can generally execute Fortran 77 with minimal user translation. Finally, we can target CM Fortran on the 544-node Connection Machine CM-5 at AHPCRC to run very large problems in its 16GB memory size [22], [20].

CM Fortran is similar to Fortran 90 and requires a coding style different from that used in the serial, vector, or shared-memory parallel machines. We can use the Fortran-P translator to convert the serial code automatically when we compile. These translation tools provide a way to maintain one code yet still run on a number of different types of machines and achieve good performance.

4.4. A TRANSLATION EXAMPLE

In this section, we describe the translation of the Miami Ocean Model code

```
      do 663 j=1,jj1
      do 663 k=2,kk
      kn=k+nn
c
      do 863 l=1,isu(j)
      do 863 i=ifu(j,l),ilu(j,l)
      q=tencm-amin1(dpu(i,j,kn),tencm)
 863  u(i,j,kn)=(u(i,j,kn)*dpu(i,j,kn)+u(i,j,kn-1)*q)/(dpu(i,j,kn)+q)
c
      do 663 l=1,isv(j)
      do 663 i=ifv(j,l),ilv(j,l)
      q=tencm-amin1(dpv(i,j,kn),tencm)
 663  v(i,j,kn)=(v(i,j,kn)*dpv(i,j,kn)+v(i,j,kn-1)*q)/(dpv(i,j,kn)+q)
```

Fig. 8. An example of the original coding style used in the University of Miami Isopycnic Ocean Model.

using the Fortran-P translator.

University of Miami Isopycnic Ocean Model. The University of Miami Isopycnic Ocean Model [2] is an oceanic general circulation model that simulates thermodynamic and wind-driven flow in realistic basins. The oceanic interior is mapped into a set of isopycnal layers giving it the performance characteristics of a "stacked shallow-water" model. The Miami model employs an explicit finite-difference technique on a staggered grid of zones to solve the appropriate PDEs.

This ocean model consists of approximately 3500 lines of Fortran 77; it was originally designed to run on vector machines such as the Cray YMP-C90. As a result, it has variables that store the positions where sections of ocean begin and end in each row and column of the horizontal domain of the problem (i.e., a layer of equal density). Each loop that computes over the entire ocean is triply nested: quite often, the two innermost loops range over a horizontal layer while the outermost loop steps vertically through the layers.

Figure 8 shows a loop from this code. The calculations done vertically (along the third or `k` index) do not parallelize and so were left as is. Because the model is "shallow," that is, the vertical dimension is much smaller than the horizontal dimensions, this is not a problem.

The outermost loop runs `j` over all the rows of the domain. The outer loop ending at `863` runs `l` over each segment in the `i` direction, while the inner loop ending at `863` runs `i` over all the elements in that segment. Thus, the code avoids doing any calculations at all over land, which may improve its efficiency on a vector machine (assuming that the vector length of *each segment* is adequate).

On a distributed-memory MPP, however, it is generally more desirable to use masks to achieve this goal by masking off the areas of land and not storing the results of calculations there. The trade-off is performing redundant calculations versus using a more complicated mapping scheme, possibly with load balancing. Fortunately, the Miami model already has mask variables (in fact, the values in the

```
      do 123 i=1,n
123   x(i)=1.0
```

Fig. 9. Do loop label before movement

```
      do 123 i=1,n
      x(i)=1.0
123   continue
```

Fig. 10. Do loop label after movement

```
      if ( y .lt. 3.0 ) goto 100
      x = 2.0
100   y = 5.0
```

Fig. 11. Goto label before movement

```
      if ( y .lt. 3.0 ) goto 100
      x = 2.0
100   continue
      y = 5.0
```

Fig. 12. Goto label after movement

isu, ifu, and ilu arrays are calculated from the mask during initialization). The **atlfix** translation phase changed the pairs of segment and index loops into single loops running over the whole domain and placed each of the statements in the loop body within an if statement with a test of the mask variable as the conditional expression. This produces equivalent Fortran 77 code that is easier to map to an MPP.[7]

Also, as the original coding style in this case did not use **continue** statements (which make the translation process easier), a small translation phase (**fixlabels**) was written that flagged all statement labels to indicate whether they were referenced by a goto statement or a do-loop. Statements labels referenced only by do-loops (Figure 9) would then be moved to a continue statement after the statement they had labeled (Figure 10). Similarly, statement labels referenced only by goto statements (Figure 11) were moved to a continue statement before the labeled statement (Figure 12). These transformations simplified the application of other translation phases.

Figure 13 shows this sample piece of code after it had been converted to the compiler intermediate form, run through the two phases constructed to modify this code (**atlfix** and **fixlabels**), and converted back to Fortran 77. All of the statement labels are now associated with continue statements, and the loop fix pass also inserted a CDIR$ serial directive so that when the code is passed through parallelization phase, the k loops will be treated as serial. Hence, the ocean basin is decomposed into columns, each mapped to a single processor; this works well since the ocean is a thin layer of fluid and the horizontal grid dimensions are generally much larger that the vertical layers. The vertical axis is Lagrangian. Note also that each statement inside an inner loop has an if statement wrapped around it so that it will only be executed at points where the mask is equal to 1.

[7] A translation of this same code for the CM-5 by R. Bleck and S. Welles at Los Alamos effectively used this same approach, i.e., masked array operations.

```
      Do 663 j=1,jj1,1
CDIR$ serial
      Do 663 k=2,kk,1
      kn=k+nn

      Do 863 i=1,ii1,1
      if (iu(i,j) .eq. 1 ) then
      q=tencm-amin1(dpu(i,j,kn),tencm)
      endif
      if (iu(i,j) .eq. 1 ) then
      u(i,j,kn)=(u(i,j,kn)*dpu(i,j,kn)+u(i,j,kn-1)*q)/(dpu(i,j,kn)+q)
      endif
863   continue

      Do 663 i=1,ii1,1
      if (iv(i,j) .eq. 1 ) then
      q=tencm-amin1(dpv(i,j,kn),tencm)
      endif
      if (iv(i,j) .eq. 1 ) then
      v(i,j,kn)=(v(i,j,kn)*dpv(i,j,kn)+v(i,j,kn-1)*q)/(dpv(i,j,kn)+q)
      endif
663   continue
```

Fig. 13. The code fragment in Figure 8 after passing through the initial translation phase

```
      subroutine stemp(x,y,n)
      integer n
      real x(n),y(n)
      real t
      integer i

      do i=1,n
         t = x(i)*x(i) - y(i)*y(i)
         y(i) = 2.0*x(i)*y(i)
         x(i) = t
      enddo

      return
      end
```

Fig. 14. A simple loop requiring scalar promotion before conversion to Fortran 90 syntax

Scalar promotion when converting Fortran 77 to CM Fortran. In the current version of the translator, we restructure codes in a number of ways so that we can express them in Fortran 90-like CM Fortran. One example of this is the scalar promotion phase. In this phase we must locate scalar temporaries that are used

```
      subroutine stemp( x, y, n)
      include 'nodes.inc'
      integer n
      real x(n)
      real y(n)
      real t(n)
      integer i
C       Begin stemp
      t(1:n)=x(1:n)*x(1:n)-y(1:n)*y(1:n)
      y(1:n)=(2.0*x(1:n))*y(1:n)
      x(1:n)=t(1:n)
      return
      end
```

Fig. 15. The loop of Figure 14 after scalar promotion and conversion to Fortran 90

only inside of loops (see Figure 14). The variable **t** is defined and used in the loop. In order to convert this code to Fortran 90 array syntax, **t** must be promoted to an array. Figure 15 shows the code of Figure 14 after passing through the scalar promotion and Fortran 90 conversion phases of the translator.

Conversion of loops to Fortran 90. The conversion of do-loops to Fortran 90 triplet syntax is done in the **cmf** phase. As this SORCERER-generated tree parser processes the tree representation of the code, it keeps a list of currently active loops. When a new loop is encountered, it is added to the list unless it was flagged as serial by a directive. When a loop ends, it is removed from the list. While the CM Fortran code is being generated for the code inside loops, array index expressions are carefully examined by a SORCERER grammar. If parallel loop indices are found in the expression, then the expression is specially processed to turn it into a Fortran 90 triplet expression. Figure 13 and Figure 16 show a loop before and after conversion to Fortran 90 syntax. Notice that the loop marked as serial (by the **CDIR$ serial** directive) was not converted to Fortran 90.

Data Layout. For the translation of the Miami model we used a domain decomposition strategy: vertical columns of the ocean were mapped to each processor. Overlapping fake zones are maintained and redundant computations performed so that minimal communication is required during the computational time step (both barotropic and baroclinic time steps) [13]. Figure 16 shows the 663 loop in Figure 13 after Fortran 90 conversion and the addition of processor dimensions; we also show the generated data layout directives.[8]

[8] In CM Fortran and HPF, data layout directives suggest to the compiler how particular array dimensions should be placed in distributed memory. In CM Fortran [22], **NEWS** indicates a dimension should be spread across processors; **SERIAL** indicates local on-processor storage.

```
      common/gindex/ iu(idm,jdm,NODE_X,NODE_Y)
CMF$LAYOUT iu(:SERIAL,:SERIAL,:NEWS,:NEWS)
      common/gindex/ iv(idm,jdm,NODE_X,NODE_Y)
CMF$LAYOUT iv(:SERIAL,:SERIAL,:NEWS,:NEWS)
      common/unnamed/ u(idm,jdm,2*kdm,NODE_X,NODE_Y)
CMF$LAYOUT u(:SERIAL,:SERIAL,:SERIAL,:NEWS,:NEWS)
      common/unnamed/ v(idm,jdm,2*kdm,NODE_X,NODE_Y)
CMF$LAYOUT v(:SERIAL,:SERIAL,:SERIAL,:NEWS,:NEWS)
      common/unnamed/ dpu(idm,jdm,2*kdm,NODE_X,NODE_Y)
CMF$LAYOUT dpu(:SERIAL,:SERIAL,:SERIAL,:NEWS,:NEWS)
      common/unnamed/ dpv(idm,jdm,2*kdm,NODE_X,NODE_Y)
CMF$LAYOUT dpv(:SERIAL,:SERIAL,:SERIAL,:NEWS,:NEWS)
C       FP-INFO: promoted scalar
      dimension q(idm,jdm,2*kdm,NODE_X,NODE_Y)
CMF$LAYOUT q(:SERIAL,:SERIAL,:SERIAL,:NEWS,:NEWS)
CDIR$ serial
      Do 663 k=2,kk,1
      kn=k+nn

      q(1:ii1,1:jj1,:,:)=MERGE(tencm-amin1(dpu(1:ii1,1:jj1,kn,:,:),tencm)
     &      ,q(1:ii1,1:jj1,:,:),iu(1:ii1,1:jj1,:,:) .eq. 1)
      u(1:ii1,1:jj1,kn,:,:)=MERGE((u(1:ii1,1:jj1,kn,:,:)*dpu(1:ii1,1:jj1
     &  ,kn,:,:)+u(1:ii1,1:jj1,kn-1,:,:)*q(1:ii1,1:jj1,:,:))/(dpu(1
     &  :ii1,1:jj1,kn,:,:)+q(1:ii1,1:jj1,:,:)),u(1:ii1,1:jj1,kn,:,:)
     &      ,iu(1:ii1,1:jj1,:,:) .eq. 1)
863   continue

      q(1:ii1,1:jj1,:,:)=MERGE(tencm-amin1(dpv(1:ii1,1:jj1,kn,:,:),tencm)
     &      ,q(1:ii1,1:jj1,:,:),iv(1:ii1,1:jj1,:,:) .eq. 1)
      v(1:ii1,1:jj1,kn,:,:)=MERGE((v(1:ii1,1:jj1,kn,:,:)*dpv(1:ii1,1:jj1
     &  ,kn,:,:)+v(1:ii1,1:jj1,kn-1,:,:)*q(1:ii1,1:jj1,:,:))/(dpv(1
     &  :ii1,1:jj1,kn,:,:)+q(1:ii1,1:jj1,:,:)),v(1:ii1,1:jj1,kn,:,:)
     &      ,iv(1:ii1,1:jj1,:,:) .eq. 1)
663   continue
```

Fig. 16. The code fragment in Figure 13 after passing through the Fortran-P translation phases. Additional parallel indices have been added, a scalar (q) has been promoted to an array, if statements have been converted to **MERGE** intrinsics, and loops have been collapsed into Fortran 90 triplet notation

Idiom Recognition. Idiom recognition [15] fits naturally within the framework of our source-to-source translator. Idioms are regular structures that can span many program statements but may contain irrelevant details; for example, inner-product calculations and recurrences, selective processing of vector elements, and repeated access patterns to distributed array sections [15]. Our high-level intermediate representation eases recognition and transformation of idioms because source program constructs are not semantically lowered towards machine language form, which

```
c#######################################################################
c
c     Calculate the buoyancy term for the w-equation.
c     The buoyancy term is stored in tem1 then added to wforce.
c
c#######################################################################
c

      tlevel =  tpresent

      CALL buoycy(nx,ny,nz, ptprt(1,1,1,tlevel),pprt(1,1,1,tlevel),
     :            qv(1,1,1,tlevel),qc(1,1,1,tlevel),qr(1,1,1,tlevel),
     :            qi(1,1,1,tlevel),qs(1,1,1,tlevel),qh(1,1,1,tlevel),
     :            ptbar,pbar,rhobar,qvbar,
     :            tem1,                                ! wbuoy
     :            tem2)
```

Fig. 17. An example of passing 3-d slices of 4-d arrays in ARPS

We first experimented with the ARPS shallow kernel code to understand some of the potential problems associated with converting the ARPS code. This code was converted in the Fall of 1992 and these early results were discussed at the ECMWF conference in November 1992 [5]. The importance of a large subgrid ratio [22] to good performance was made clear by these early studies.

Our attention shifted during the winter of 1993 to conversion of the full ARPS code, in particular, the ARPS timestep kernel. This kernel code conforms directly to the current Fortran-P model, which generally excludes initialization code. Our first working version used the original scalar initialization code and performed all the initialization on the CM-5 front-end processor [7]. The initialized state arrays were then sent to the processing nodes. A similar technique was employed in order to use the I/O routines without modification: the data on the nodes was transferred back into a front-end array and passed to the original scalar I/O routines. Front-end memory size and processing speed limited the problem size to $80 \times 80 \times 80$ grid cells, so we later rewrote the initialization and I/O routines in CM-Fortran in order to execute larger problem sizes.

Difficulties with CM-Fortran. The version of the CM-Fortran compiler we used for this program was CMF 2.1 Beta 0.1. We encountered several difficulties related to this version of the compiler and/or libraries.

1. Compiler generated temporaries were unreliable in some cases. The ARPS model has 4-d arrays containing the model state over three space dimensions and one time dimension. In several places, routines are passed a 3-d slice of a 4-d array, which corresponds to passing the state at a particular time. An example of the coding style is shown in Figure 17.
   By replacing the `1`'s with `:`'s this becomes acceptable CM-Fortran code. Unfortunately, there seemed to be some problem with the way that the compiler dealt with this. With the `prism` debugger we could see that the 3-d slice was correct

might tend to obscure common idioms. For example, when translating the Miami model, we replaced `if` statements with single assignment statements in the `then` and `else` clauses ("`MERGE`able" `ifs`) with the Fortran 90 `MERGE` intrinsic, which is much faster on the CM-5. The elimination of `if` statements inside of parallel loops is also necessary because they would prevent the `cmf` phase from converting the loops to Fortran 90 triplet form. An example code sequence before and after translation is shown in Figure 13 and Figure 16, respectively.

4.5. CURRENT STATUS OF THE MIAMI OCEAN MODEL TRANSLATION

Our translation of the Miami ocean model is largely complete at the time of this writing. The code's principle author, Dr. Rainer Bleck, has assisted the translation process by modifying the boundary treatments in the Miami model to conform to the Fortran-P constraints. At the time of this writing we are currently debugging the Fortran-P version of the code. Additional work remains in converting the I/O and initialization sections of this code; these are relatively small and do not have a significant performance impact, so they will be translated by hand.

A complete hand-translation of the Miami model into CM Fortran by S. Welles at Los Alamos National Laboratory[9] took approximately one year [3]. Not including the time to construct the reusable parts of the translator, the conversion time using our tool-based approach will require about one man-month. In addition, our restructured Fortran-P version of this code will easily port to message-passing form; we are currently working with Cray Research to port this code to the Cray MPP using `get/put` message-passing primitives.

5. Conversion of the ARPS 3.0 Mesoscale Weather Code

In this section we describe how the Fortran-P translator assisted in the translation of a meso-scale weather code. This semi-automatic translation was completed in the early summer of 1993 and experience has been gained actually executing this code on the CM-5 using large problem domains.

The ARPS is a meso-storm-scale prediction Fortran 77 code under development by the Center for Analysis and Prediction of Storms (CAPS) at the University of Oklahoma. ARPS is a 3-D, fully compressible, nonhydrostatic weather model that includes cloud physics (humidity, cloud water, and rain water), surface effects, and subgrid-scale turbulent mixing. The ARPS uses second-order quadratically-conservative spatial discretization and second-order leapfrog (with Asselin time filter option) temporal discretization, based on an Arakawa C-grid with terrain-following vertical coordinates. The code for the model is over 50,000 lines of modular Fortran-77 [7].

With the converted ARPS code we simulated the 20 May 1977 Del City, Oklahoma storm, a well-documented tornadic supercell. The ARPS source code includes sounding and input files to carry out a one-hour storm simulation [7].

5.1. THE INITIAL CONVERSION OF ARPS 3.0

[9] This translation did not use a domain decomposition strategy and converted the horizontal axes directly to Fortran 90 parallel.

in the routine called, but the data values became corrupted one or two levels farther down the call tree. We defined explicit 3-d temporaries and copied the appropriate 3-d slices of 4-d arrays to them before calling routines that needed 3-d inputs.

2. The CM-Fortran compiler we employed for the ARPS conversion (CMF 2.1 Beta 0.1) would not vectorize across serial dimensions [5]. The Fortran-P group gained the attention of the Thinking Machines Fortran group and requested that this constraint be removed, which in fact has happened in the current release of CM Fortran (CMF 2.1 Beta 1.0) [21].[10] As a result of this constraint we needed high VP ratios [22] in order to get good performance. This in turn meant having many subdomains assigned to each processor. Since the subdomains are similar to the original problem domain and since ARPS uses 3 boundary points (one on the left and two on the right) each subdomain also needed to have 3 boundary points. Since ARPS uses a lot of memory (it has about 50 full-domain-sized 3-d arrays as well as full-domain-sized temporaries) the subdomains needed to be small to avoid running out of memory on the CM-5. A reasonable size turned out to be 5^3 real points per subdomain, which means 8^3 total points per subdomain. This means that only $5^3/8^3 = 24.4\%$ of the memory used was used for real points, the remainder being used for boundary zones.

   This memory inefficiency when using large VP ratios resulted in ARPS computing at about 1/4 the speed of actual weather. This speed was observed on 32-nodes of the 544-node CM-5 at the U. of Minnesota using a grid size of $40 \times 40 \times 40$ (grid point resolution of 1km). For a 128-node run a grid size of $80 \times 80 \times 40$ ran at about the same speed as the actual weather.

   In Table 2 some problem size, subdomain size, processor grid, and memory usage figures are given. The CM-Fortran compiler has what is called the "rule of 8" which says that array extents will be padded to the next multiple of 8.[11] This is why the $32 \times 32 \times 32$ problem takes so much more memory than the $40 \times 40 \times 40$ problem: with the 3 boundary zones used by ARPS, its subdomain arrays are $11 \times 11 \times 11$ and were presumedly padded to be $16 \times 16 \times 16$. Also, the table shows both the ratio of subdomains to processor nodes, known as the Virtual Processor (or VP) ratio, and the ratio of subdomains to vector nodes, which we refer to as the Vector Virtual Processor (or VVP) ratio, which is different since each CM-5 processor node has 4 vector processors. The latter ratio is referred to by TMC as the *subgrid ratio*.

3. While debugging and executing the converted ARPS code, we found there were some problems in the converted versions of the turbulent mixing code. These problems often became evident later in the simulations when several variables would become unstable. The turbulent mixing code required somewhat involved boundary handling code, much of which had to be generated by hand given the constraints of the α version Fortran-P translator used at that time. (These

[10] Many thanks to Woody Lichtenstein and others at Thinking Machines for their help on this issue and others related to CM Fortran.

[11] The "rule of 8" has also been relaxed in CMF 2.1 Beta 1.0; there is now a **-nopad** option that turns off this padding of array sizes.

Problem Size	Subdomain	Proc. Grid	Proc.	VP	subgrid	Memory
$32 \times 32 \times 32$	$8 \times 8 \times 8$	$4 \times 4 \times 4$	32	2	0.5	578MB
$40 \times 40 \times 40$	$5 \times 5 \times 5$	$8 \times 8 \times 8$	32	16	4	243MB
$80 \times 80 \times 80$	$5 \times 5 \times 5$	$16 \times 16 \times 16$	128	32	8	977MB

TABLE I
Problem size, subdomain size, processor grid size, and resulting VP and subgrid ratios

constraints are being lifted in the β version Fortran-P translator currently under test at Minnesota.) These difficulties emphasized the importance and power of automatic, correct-by-construction translation tools.

4. Long compiles were a bottleneck: it was not unusual for compiles and links to take up to an hour, making it much more difficult to get through the edit-compile-debug cycle. Even re-compiling and linking a few files would often take 10–20 minutes. This problem seems to be slowly improving with new releases of the compiler.

Preliminary Results with CMF 2.1 Beta 1.0 and β Fortran-P. As mentioned earlier, CM Fortran compiler restrictions for versions prior to CMF 2.1 Beta 1.0 (which we unfortunately could not use in this preliminary study) yielded poor memory efficiency and hence poor computational efficiency as well given the Fortran-P model of converted code. The α version Fortran-P translator did not generate the appropriate loop form for the CM Fortran 2.1 Beta 1.0 compiler; however, our early β Fortran-P translator was used to translate parts of the ARPS code and study its performance. These early results with the CMF 2.1 Beta 1.0 compiler and the new Fortran-P translator have been very promising. In particular, high subgrid ratios become much less important and replication of subgrids on processors (which leads directly to memory inefficiency) was not critical to performance. In addition, the **-nopad** compiler switch prevents padding of arrays, also leading to better memory efficiency. We have also employed an early version of the β Fortran-P translator to generate Fortran-90-style loops (instead of the serial DO loop structure employed in the α version) and used the Cray FPP inliner to reduce the significant subroutine overheads present in ARPS due to its operator subroutine style [7]. These translation improvements led to significant performance increases; early results indicate that the ARPS code at $40 \times 40 \times 40$ resolution now runs at two times the speed of the weather on 32-nodes (a factor of 8 increase in speed over our earlier results).

Data I/O and Visualization. We used the CM-5 DataVault to save dumps of the model state at intervals during the run. At the largest resolution we executed and visualized the code ($160 \times 160 \times 80$) each dump of one variable required 8MB of disk space (the data were stored as 32-bit floating point numbers). We dumped 7 variables: the u, v, and w velocity fields, perturbation pressure, perturbation temperature, cloud water, and rain water. Thus we required 56MB per data dump, or about 3.4GB to do a dump every 60 seconds during a 1 hour simulation. After the run was done, a second program was used to read the dumps in 32-bit float format, scale them down to one 8-bit byte per data point, and dump them on the CM-5 front end in Brick-Of-Bytes (BOB) format This scaling generally took about 5 seconds per frame if the CM-5 was not busy with other jobs. We discovered that running two copies of the scaling program on the same CM-5 partition would crash the DataVault (the CM-5's disk subsystem), but multiple copies on different partitions did not seem to cause problems. The resulting BOB files were then visualized with the AHPCRC Volume Renderer on the SGI workstations [4]. The rendered frames were stored on an animation drive [17] and then played back with the **xraz** raster animation tool. A short video of these animations is available in VHS format upon request (send requests to **okeefe@ee.umn.edu**).

5.2. SUMMARY OF ARPS TRANSLATION AND EXECUTION

In this section we have detailed our translation of the ARPS 3.0 meso-scale weather code and described its performance on the CM-5 at the University of Minnesota. Though current performance results are lower than we had hoped, the performance problems have obvious solutions; our preliminary results using new compiler and translator software are promising. We intend to re-translate ARPS 3.2 and study its performance as well as perform some very high resolution supercell storm simulations on the CM-5.

Our experience with the ARPS code has pointed out many important properties of other finite difference self-similar codes, including the use of staggered grids. We are incorporating new features in our translator to reflect this new understanding obtained through the ARPS code translation effort.

6. Conclusions and Future Work

In this paper, we have described the construction of a source-to-source translator from Fortran 77 (Fortran-P) to CM Fortran and its application to the semi-automatic conversion of the Miami ocean model and the ARPS. We described our tool-based approach to source-to-source translation via independent phases that may be reused when constructing other translators and compilers. With the inclusion of phases to compute a dependence or computation graph [15], more analysis and sophisticated global translation can be incorporated, providing an evolutionary path to more powerful translation. We intend to incorporate these phases as the translator evolves and we apply it to optimize and parallelize more codes.

We have also applied the translator tools to assist in the conversion of several other codes including a 5,000-line hydrodynamics code used to model compressible, turbulent flows [25], [13]. These codes have achieved good performance on very large grids using the large memory available on the CM-5. Our goal is to continue to refine and develop our current translator, applying it to more codes and using it to measure the performance attainable on a variety of high performance computers, including massively parallel processors and workstations. We are currently retargeting our tools to generate code for the Cray MPP and clusters of Silicon Graphics Challenge servers. Additional transformation phases are being added to optimize memory usage for very large grids and to automate boundary handling and updating.

References

[1] T. Blank, "The MasPar MP-1 Architecture," 35th IEEE Computer Society International Conference (COMPCON), pp. 20-24, February 1990.

[2] R. Bleck, C. Rooth, D. Hu, and L. Smith, "Salinity-driven Thermocline Transients in a Wind- and Thermohaline-forced Isopycnic Coordinate Model of the North Atlantic," *J. of Physical Oceanography*, vol. 22, no. 12, pp. 1486–1487, December 1992.

[3] R. Bleck, *Personal communication*, September 1993.

[4] K. Chin-Purcell, "All About BOB: A Tool for Browsing 3D Data Sets," Army High Performance Computing Research Center Preprint #92-141, University of Minnesota, December 1992.

[5] K. Droegemeier, K. Johnson, K. Mills, and M. O'Keefe, "Experiences with the Scalable-Parallel ARPS Cloud/Mesoscale Prediction Model on Massively Parallel and Workstation Cluster Architectures," *Proc. 5th Workshop on the Use of Parallel Processors in Meteorology*, European Centre for Medium Range Weather Forecasts, Reading, England, November 1992.

[6] K. Droegemeier, M. Xue, K. Johnson, A. Sawdey, M. O'Keefe, *et al.*, "The Design and Implementation of a Scalable-Parallel Stormscale Numerical Weather Prediction Model," to appear in *High Performance Computing*, Addison-Wesley Publishers, ed. by G. Sabot *et al.*, 1994.

[7] Center for Prediction and Analysis of Storms. *ARPS Version 3.0 User's Guide.* Center for Analysis and Prediction of Storms, Norman, Oklahoma, 1992.

[8] M. Gerndt, "Updating Distributed Variables in Local Computations," *Concurrency: Practice and Experience,* vol. 2, no. 3, September 1990, pp. 171–193.

[9] M. Gupta and P. Banerjee, "Demonstration of Automatic Data Partitioning Techniques for Parallelizing Compilers on Multicomputers," *IEEE Trans. on Parallel and Distributed Systems,* vol. 3, no. 2, pp. 179–193, March 1992.

[10] High Performance Fortran Language Specification, *High Performance Fortran Forum*, May 1993.

[11] O. Lubeck, M. Simmons, and H. Wasserman, "The Performance Realities of Massively Parallel Processors," *Supercomputing '92,* pp. 403–412, Minneapolis, MN, November 1992.

[12] W. Noh and P. Woodward, "SLIC (Simple Line Interface Calculation)," *Lecture Notes in Physics*, vol. 59, 1976.

[13] M. O'Keefe, T. Parr, B. K. Edgar, S. Anderson, P. Woodward, and H. Dietz, "The Fortran-P Translator: Automatic Translation of Fortran 77 Programs for Massively Parallel Processors," Army High Performance Computing Research Center Preprint #93-021, University of Minnesota, February 1993. Submitted to *The Journal of Scientific Programming.*

[14] T.J. Parr, "SORCERER—A Source-to-Source Translator Generator," Army High Performance Computing Research Center Preprint #93-088, University of Minnesota, September 1993, submitted to the 1994 *International Conference on Compiler Construction.*

[15] S.S. Pinter and R.Y. Pinter, "Program Optimization and Parallelization Using Idioms," *Proc. 1990 Symposium on the Principles of Programming Languages*, pp. 79–92.

[16] D. Porter, A. Pouquet, and P. Woodward, "Kolmogorov-Like Spectra in Decaying Three-Dimensional Supersonic Flows", submitted to *Physics of Fluids.*

[17] T. Ruwart, "High-Speed, Disk-Based Interactive Scientific Animation," *Supercomputing Review,* September 1991.

[18] A. Sawdey, M. O'Keefe, O Meirhaeghe, M. Xue, and K. Droegemeier, Conversion of the ARPS 3.0 Meso-Scale Weather Prediction code to CM-Fortran using the Fortran-P Translator, Army High Performance Computing Research Center Preprint #93-089, University of Minnesota, August 1993.

[19] Thinking Machines Corp., *Connection Machine Model CM-2 Technical Summary,* TMC Technical Report TR89-1.

[20] Thinking Machines Corporation, *Connection Machine Model CM-5 Reference Manual.* Cambridge, Massachusetts, 1991.

[21] Thinking Machines Corporation, *CM Fortran Version 2.1 Release Notes.* Cambridge, Massachusetts, June 1993.

[22] *CM Fortran Reference Manual.* Thinking Machines Corporation, Cambridge, Massachusetts, 1991.

[23] M. Walker, "The Cray MPP," talk given at *ECMWF Conference on Parallel Processors in Meteorology*, Reading, England, November 1992.

[24] M. J. Wolfe, *Optimizing Supercompilers for Supercomputers,* Cambridge, MA: MIT Press, 1989.

[25] Paul R. Woodward, "Piecewise-Parabolic Methods for Astrophysical Fluid Dynamics," *Astrophysical Radiation Hydrodynamics*, pp 245–326, D. Reidel Publishing Company, ed. by K.H.A. Winkler and M.L. Norman, 1986.

GLOBAL ATMOSPHERIC AND OCEAN MODELING ON THE CONNECTION MACHINE

SUSAN R. ATLAS*
Thinking Machines Corporation
245 First St.
Cambridge, MA 02142 USA

ABSTRACT. This paper describes the high-level architecture of two parallel global climate models: an atmospheric model based on the Geophysical Fluid Dynamics Laboratory (GFDL) SKYHI model, and an ocean model descended from the Bryan-Cox-Semtner ocean general circulation model. These parallel models are being developed as part of a long-term research collaboration between Los Alamos National Laboratory (LANL) and the GFDL. The goal of this collaboration is to develop parallel global climate models which are modular in structure, portable across a wide variety of machine architectures and programming paradigms, and provide an appropriate starting point for a fully coupled model. Several design considerations have emerged as central to achieving these goals. These include the expression of the models in terms of mathematical primitives such as stencil operators, to facilitate performance optimization on different computational platforms; the isolation of communication from computation to allow flexible implementation of a single code under message-passing or data parallel programming paradigms; and judicious memory management to achieve modularity without memory explosion costs.

1. Introduction

As an enterprise of inherently international scope, the long-term, high-resolution simulation of the Earth's global climate presents unique opportunities for large-scale collaboration and research. The prospect of imminent availability of raw processing power in the 100 GFlops range, combined with rapid and continuing refinements in the models themselves, led to the initiation of the U. S. Department of Energy CHAMMP (Computer Hardware, Advanced Mathematics, and Model Physics) Climate Modeling Program in 1990 [1]. The mission of this program is to develop new algorithms and numerical techniques to take advantage of the increasing computational power provided by a new generation of (parallel) computer architectures, and thus to facilitate the development of integrated models that incorporate the effects of such complex processes as atmospheric photochemistry, land vegetation, and atmosphere/ocean/sea-ice interactions.

---

* The work described in this paper is the result of a collaboration between scientists at Los Alamos National Laboratory (LANL), the Geophysical Fluid Dynamics Laboratory (GFDL/NOAA, Princeton University), Thinking Machines Corporation (TMC), and Cray Research, Inc. (CRI), under the auspices of the U. S. Department of Energy CHAMMP (Computer Hardware, Advanced Mathematics, and Model Physics) Climate Modeling Program, directed by R. C. Malone (LANL). The collaboration consists of J. K. Dukowicz, P. W. Jones, S. Kortas, R. C. Malone, and R. D. Smith (LANL); C. H. Goldberg and R. Hemler (GFDL); S. R. Atlas and K.-S. Cho (TMC); and C. L. Kerr (CRI).

F.-X. Le Dimet (ed.), High Performance Computing in the Geosciences, 245–255.

For the past several years, the Climate Modeling Group at Los Alamos National Laboratory (LANL) has been engaged in a research collaboration with the Geophysical Fluid Dynamics Laboratory of the National Oceanic and Atmospheric Administration (GFDL/NOAA), under the auspices of CHAMMP, to develop massively parallel versions of two of the GFDL's ocean and atmospheric modeling codes. The initial target for the parallel versions of these codes has been the Connection Machine installed at LANL's Advanced Computing Laboratory (ACL)– initially a 64k CM-200, which was recently upgraded to a 1024 node CM-5. Since the CM-5 can be programmed to run in either message passing or data parallel mode [2], it has been possible to address the issue of portability concurrently with the design of the parallel models themselves. The eventual goal is to produce full-scale models which can run on virtually any platform, ranging from a serial, standalone workstation, to a multitasked Cray YMP/8, to a loosely-coupled cluster of workstations running PVM, to a massively parallel machine such as the CM-5 running High Performance Fortran (HPF)– *all without significant sacrifice of performance on any individual platform.* This is particularly important for implementations running on the largest parallel machines, such as the CM-5, since these will be used to perform very long, high-resolution experiments, probing the limits of climate predictability on centuries-long timescales.

Another driving force in this work, particularly with regard to the atmospheric model, has been the development of modular codes, to facilitate the interchange of individual components such as radiation packages or dynamics solvers within the models. In this way, different implementations (grid vs. spectral; alternate physical parametrizations) can be compared in a controlled fashion, or provided to researchers as standalone modules.

The focus of this paper will be the high-level architecture of the ocean and atmosphere models and their implementation on the CM-5. We begin with a history and brief description of the atmospheric and ocean models in Section 2, followed by an overview of the CM-5 architecture in Section 3. The design of the parallel models is described in Section 4, with examples to illustrate the accomodation of various communication, memory management, and programming paradigms. We conclude with a discussion of future directions in Section 5.

2. Two Models

2.1 SKYHI

SKYHI is a general circulation model of the atmosphere, developed at GFDL [3]-[6]. It is a global hydrostatic model with 40 levels extending to 80 kilometers, utilizing a rectangular latitude/longitude grid in the horizontal, and a sigma coordinate system in the vertical. Fourier filtering is used to eliminate high-frequency features near the poles. The model integrates the primitive equations for the prognostic variables, including wind velocity, temperature, water vapor mixing ratio, and (optional) user-selected tracers. Radiative heating, an input to the primitive equations, is re-calculated every fixed number of timesteps, as specified by the user. Following each time step, diagnostic variables are calculated and optionally written out to storage.

The process of converting SKYHI to parallel form began in 1990, with the development of an array-syntax version of the original 40,000 line Fortran 77 code. This new version of SKYHI consisted of 20,000 lines of Fortran 90 which could be run on GFDL's Cray YMP/8 following the application of a pre-processor [7] that converted the Fortran 90 back into Fortran 77 for the Cray compiler. At that point, the code was still structured in terms of array "chunks", in which prognostic variables were integrated in successive grid slices, each three latitudes wide. This was necessary because of the limited amount of main memory on the Cray. In November, 1992, work began at LANL and GFDL to further evolve SKYHI into a true parallel model. An "in-core" model was developed, full conversion to array syntax was completed, and HPF-style layout directives were added to specify domain decomposition across processors. The new, parallel version of SKYHI is presently undergoing performance optimization and verification on the CM-5 at LANL and on the YMP/8 at GFDL. The first numerical experiments using SKYHI on the CM-5 will begin in early 1994.

2.2 PARALLEL OCEAN PROGRAM (POP)

POP is a global ocean model presently running in production on the CM-5 at LANL. It was originally developed as a CM-2 port [8] of the Semtner-Chervin ocean model [9,10]. The Semtner-Chervin model is a descendant of the Bryan-Cox model [11], variants of which form the basis for many ocean modeling programs in use today, including the MOM model from GFDL. As detailed in reference [8], the port of the Semtner-Chervin model to the CM-2 required a significant re-write, since, as in SKYHI, the data structures used in the Cray implementation were ill-suited to the efficient use of a massively parallel architecture. The authors of the port (Smith et al.) also found that the performance of POP on the CM-2 was severely limited by the barotropic solver implementation in the original code. They subsequently developed a reformulation of the solver in terms of surface-pressure (rather than the streamfunction) which not only parallelized better on the CM-2 but also made it possible to include an arbitrary number of islands in the model without excessive computational cost [12,13]. For further details, see [8].

Many of the programming features which promote modularity and portability, such as isolating communication within stencil routines (see Section 4.3) were first developed in POP. The model now runs on a variety of architectures, including a workstation cluster running PVM, Cray vector computers, and several massively parallel supercomputers, including the CM-5. As part of the LANL/GFDL collaboration, the capabilities of the model are presently being expanded to include the complete range of diagnostics and user interface available in MOM.

3. CM-5 Architecture

The CM-5 system [2] consists of a set of 16 to 16,384 processing nodes (PNs) with associated memory. Each PN contains a SPARC microprocessor, 32 to 128 MB of memory, four vector execution/memory interface units, and a network interface. Peak processing power is 32 MFlops per vector unit, for a peak of 128 MFlops/PN. The processing nodes are coordinated by a control processor (CP), essentially a SPARC microprocessor. The CM-5 can be divided (by the system administrator) into groups of processors called partitions. Each partition is under the control of its own CP, which runs an extended form of Unix known as CMOST. The CP is responsible for coordinating and timesharing the processes in its partition. The PNs and CPs are connected to two scalable networks, the Data Network and the Control Network, both of which are based on a hyper-tree topology. The Data Network supports point-to-point communication between processors, for example SENDs and RECEIVEs in MIMD-style message passing. The structure of the Data Network avoids bandwidth contention between partitions, and the design guarantees a minimum network bandwidth of 5 MB/sec per node, regardless of destination. The Control Network supports communication patterns that involve all processors in a single operation; for example, broadcasts, synchronization, and error signaling. The Control Network also provides hardware support for cooperative arithmetic operations such as parallel prefix and reduction.

Parallel I/O to the CM-5 is provided by the Scalable Disk Array (SDA), a high performance, scalable, RAID 3 disk storage system. Like the PNs, the SDA disk storage nodes are connected to the CM-5 Data Network through a network interface. Each SDA module consists of three disk storage nodes, and provides 25 Gbytes of storage and 33 MB/s real application bandwidth. Storage capacity and bandwidth scale linearly with the number of modules. Data can be transferred to and from the SDA *in parallel* using standard Fortran READs and WRITEs, and SDA files are both UNIX compatible and NFS mountable.

4. Code Design

4.1 'GENERIC' PARALLEL COMPUTATION

At the simplest level, a parallel computer can be viewed as a collection of processing nodes, or PN's, with associated memory, coupled by a high-performance network. For purposes of high-level code design, the actual PN architecture and network topology are largely irrelevant. The fundamental question which determines the structure of a given applications program is the *domain decomposition*: the distribution of the data across the processors which are to operate upon it. Once this crucial decision has been made, it is possible to implement any given algorithm, across a wide variety of machines, using a single high-level program.

There is a fundamental reality of parallel computing which makes this statement true. This is the fact that all parallel machines in existence today, including the CM-5, are communcation- rather than computation-limited. Thus, there is no case in which it is desirable to structure the decomposition of a problem to avoid computation by shuffling data within the machine.

Once the domain decomposition is fixed, it is the responsibility of the programmer to ensure that the results of the computation that each PN has performed on its data are available to any other processors that might require it during the next round of computation. The way in which this is done is architecture-dependent, and is determined by the degree of control that the user wishes to retain over communication. This choice, in turn, is guided by the structure of the memory hierarchy on the machine, and the flexibility of the programming languages and compilers which manage that memory within an application.

As a concrete example, consider a distributed-memory parallel computer such as the CM-5. There are two programming styles that can be used to program the CM-5: data parallel and message passing. In data parallel programming, Fortran 90 array syntax [14] is used to expose the inherent parallelism of a data set. For example, in both SKYHI and POP, it forces the programmer to deal with the full latitude/longitude/depth grid as a single data structure, rather than as individual longitude/depth slabs (as in the original Cray Fortran 77 versions). However, Fortran 90 syntax says nothing about the manner in which data is to be laid out across processors on an actual machine. On the CM-5, this *domain decomposition* is accomplished through the addition of HPF-style directives [15], which give the user considerable control over the precise arrangement of data on each PN. For example, consider an array of the form A(i,j,k), where (i,j) labels a point on the latitude/longitude grid, and k is the vertical coordinate. This is a common construct in SKYHI. Using CM Fortran LAYOUT directives (DISTRIBUTE directives in HPF) one can specify that the (i,j) axes be spread across all the processors on the machine, and that the k axis be located entirely in-processor. This is written as:

```
real, array(imax,jmax,kmax) ::  A      ! F90 array syntax
cmf$ layout A(:news, :news, :serial)   ! domain decomposition
```

When this array is used in calculations involving other arrays of the same size and LAYOUT (such arrays are said to be "conformable"), operations involving variables that correspond to the same grid point but different vertical index values (common in the radiation package, for example) will involve no communication between processors. Note that there is no need for the programmer to have any *a priori* knowledge about the size of the machine on which the code is eventually to run; the actual data distribution occurs at run time, and information on the location of each grid point is stored in the *geometry* associated with the array. During compilation, the LAYOUT information is used to insert low-level communication routines into the object code wherever the compiler detects the possibility of off-processor data motion. Depending on the fine-grainedness of the control exercised by the programmer, there may be instances where the compiler inserts communication calls "just to be safe", but in fact, information is really only being moved within the memory of the processor itself. One way of avoiding this unnecessary "communication" is to assert full control over the data distribution, and insert inter-processor communication calls from within the high-level program itself. This can can be done using a message-passing programming model. Another way is to use detailed memory management, a form of local array equivalencing in CM Fortran. In general, data parallel codes tend to be faster to write, easier to maintain, and more straightforward to understand than message-passing code (the correspondence between code and equations is more transparent), but to achieve performance comparable to a tailored message-passing code, careful attention must be paid to memory management and the minimization of unnecessary communication.

4.2 PROGRAMMING MODELS

As mentioned in the Introduction, an important part of this work was the design of a high-level structure for the two models which could provide a unified framework to support a range of programming styles. Since the CM-5 supports both message passing and data parallel programming paradigms, the availability of such a framework incidentally provides the interesting opportunity to make a direct comparison between two different implementations of the same code on the same machine (this work is in progress for POP). While such a comparison will undoubtedly provide useful information, it is important to emphasize that the main reason for undertaking the development of a unified framework is to allow a single version of each model to be run on a wide variety of platforms, under a range of compilers.

An important early decision in the development of this framework was the adoption of Fortran 90 as the "maintenance language" for both models. This language has recently become an international standard, intended as the eventual replacement for Fortran 77. Until Fortran 90 compilers become widely available on serial workstations and vector supercomputers, however, a pre-compiler can be used to transform Fortran 90 syntax into Fortran 77.

Another design decision was the adoption of HPF syntax to describe the distribution of data across processing nodes. Since HPF is not yet a standard, a mechanism for supporting vendor-specific versions of these directives had to be developed. We have utilized the simple expedient of a "definitions file" #included in each subroutine; this file contains vendor-specific #ifdefs which are turned on or off for appropriate pre-processing under cpp (C pre-processor) in an architecture-specific makefile. Here is the generic form of the CM-5 layout example from Section 4.1:

```
File define.h:

#ifdef distributed_memory
   #define space_                ! blank space requires special handling
   #define hpf_distribute cmf$ space_ layout
   #define block_  :news
   #define star_   :serial
#endif

SKYHI subroutine:

   subroutine example
#include "define.h"
   real, array(imax,jmax,kmax) :: A        ! F90 array syntax
#ifdef distributed_memory
   hpf_distribute A(block_, block_ ,star_) ! HPF domain decomposition
#endif
   return
   end
```

The actual "multi-purpose programming structure" [16] for POP and SKYHI was developed by R. C. Malone and C. L. Kerr at GFDL during the summer of 1991 [16,17]. It utilizes self-similar domain decomposition, coupled with encapsulated communication. A summary of the approach is as follows. (Note: for simplicity, we shall ignore the vertical coordinate in this discussion; as described in Section 4.1, in atmospheric and ocean modeling codes this coordinate is naturally "serial", i.e. in-processor. The domain decomposition problem is thus reduced to a consideration of the distribution of data with respect to latitude/longitude coordinates). Consider a grid *NX* x *NY* in extent. Define a finer level of granularity, a *subgrid* of the larger grid, dimensioned *nx* x *ny*, and let $nx = NX / xparts$; $ny = NY / yparts$. Let the code be structured so that all mathematical operations involving communication (in POP and SKYHI these are either stencil operations corresponding to differential operators, or global sums to compute diagnostics) are encapsulated into subroutines that are dimensioned at the size of the subgrid. Use makefiles to ensure that the appropriate (depending on the programming model) versions of each communication-based subroutine are linked in to the rest of the program (including a manual domain-decomposition subroutine invoked at the start of the program in the case of message-passing). Then, as detailed in [17], it is possible, with this high-level structure, to support both data parallel and message-passing implementations on distributed memory machines, as well as "worksharing" [18] on distributed or shared memory architectures. The implementation is determined by the combination of communications routines and the values selected for *xparts* and *yparts*. For example, consider the case of a distributed-memory parallel machine such as the CM-5. When *xparts*=*yparts*=1, the "subgrid" extends over the entire global domain; the code is an ordinary data parallel program, and the corresponding communications-based subroutines are expressed in terms of Fortran 90 constructs such as CSHIFTs. The actual communication is handled implicitly by the CM-5 run-time system. When *xparts* and *yparts* are > 1, the subgrids are local to the processing nodes, and the communications-based subroutines employ MIMD-style SENDs and RECEIVEs. In this latter case, the availability of Fortran 90-to-Fortran 77 preprocessors makes it possible to support message-passing architectures other than the CM-5 (eg., the Intel Paragon) for which node-level Fortran 90 compilers are not yet available.

4.3 MATHEMATICAL PRIMITIVES

In optimizing the data parallel version of POP on the CM-5, immediately following its migration from the CM-200 (late 1992), it was determined that stencil operations on the CM-5 were occupying a fairly large percentage of the total execution time. The two main stencils involved were a nine point stencil in the barotropic solver, and a five point stencil in the baroclinic solver. At very large subgrid ratios (~ 2048) [19], these operators were responsible for over 33% of the run time; at a moderate subgrid size of 128 (a size more typical of anticipated production runs of POP at LANL), this figure was roughly 72%. A detailed analysis of these results [20] showed that by optimizing communication at the level of the run-time-system (RTS), performance could be improved significantly. This was accomplished by adding generalized support within the RTS to cache geometry information, making it possible for the CM-5 to perform the arithmetic component of the stencil operations with a minimal amount of communications overhead.

In addition, since POP had been written in such a way as to isolate the communications component of the code (i.e. the stencils), it seemed natural to try to optimize their performance directly, at the assembly-language level. A summary of results obtained using these hand-coded stencils (written by E. D. Dahl of Thinking Machines Corporation) illustrates the improvement that was obtained, particularly at small subgrids (Table 1):

Table 1. Five point stencil performance on the CM-5.
"Stencil 1" denotes stencils coded using data parallel Fortran/RTS without geometry caching. "Stencil 2" denotes hand-coded assembly language stencils. All stencil times in seconds, double precision calculations, 20 vertical levels.

Grid	Partition size (PNs)	Subgrid ratio	Stencil 1	Stencil 2
256 x 128	1024	8	3.172	0.605
128 x 64	128	16	3.165	0.681
256 x 128	512	16	3.220	0.700
256 x 128	256	32	3.278	0.776
512 x 256	1024	32	3.367	0.801
256 x 128	128	64	3.453	0.993
512 x 256	512	64	3.485	1.014
512 x 256	256	128	4.085	1.231
1024 x 512	1024	128	3.962	1.237
512 x 256	128	256	5.425	1.731
1024 x 512	512	256	5.250	1.827

Note that asymptotically, timings are determined entirely by the subgrid ratio; i.e., two different grids run on two different machines will execute in roughly the same time if the subgrids for the two problems are the same. At small subgrid sizes, the stencil time is dominated by the start-up time (latency) for off-node communication; as the subgrid increases, the startup time is amortized over the time required to actually move the data. (N.B. The version of the RTS which caches geometry information yields similar results).

In addition to illustrating the isolation of communication from computation (Section 4.1) in a parallel climate code, the use of stencils in POP serves to demonstrate the value of defining high-level mathematical primitives in scientific applications [21]. In this context, "high-level primitives" refer to mathematical operators with associated physical significance– something beyond the level of a simple Ax + B operation, for example. Programming in terms of such primitives makes sense not only because it produces clearer code, but also because it provides a natural bridge between applications scientists and software engineers. Language and compiler development for parallel machines presents significant challenges, and identification of the "natural" primitives arising in real scientific applications provides valuable feedback which can influence the future directions of compiler design.

4.4 MEMORY MANAGEMENT

As indicated in the Introduction, it is difficult to separate issues of memory management from communication when dealing with parallel machines. The relative magnitudes of memory access times within the memory hierarchy of a given machine lead to architecture-dependent compiler tradeoffs, typically within the context of communication-related constructs. These hidden tradeoffs are inevitable, and they represent the next major challenge in proceeding with the development of efficient, unified, parallel versions of the two global climate models described in this paper. In Section 4.2 it was shown that encapsulated communication provided a solution for dealing with the very different communication protocols of the data-parallel and message-passing programming paradigms. It also made it possible to focus on performance issues within a small, carefully-circumscribed domain. However, it now seems clear that the unifying framework of Section 4.1 will need to be extended to address the more complex issues associated with efficient memory management across a range of computational platforms.

As a simple example, consider the following tradeoff between the new "serial optimizations" in Version 2.0 of the (data parallel) CM Fortran compiler, and the memory explosion which can occur when the additional arrays required to implement these optimizations are introduced. In the new version of the compiler, array section assignments along serial dimensions are recognized as being in-processor, and generate local memory references instead of RTS communication calls. This makes it possible to write code that allows the compiler to vectorize along both parallel (subgrid) and serial axes at the same time, leading to significant performance enhancements. The tradeoff is memory, as illustrated in the following hypothetical fragment from an atmospheric model radiation package:

<u>Version 1 (2D arrays, no serial optimization):</u>

```
      real, array(imax, jmax) :: A, B      ! latitude/longitude grid
cmf$ layout A(:news, :news), B(:news, :news)
      do k=2,kmax                          ! unrolled DO loop over serial axis
        call physics_1(A,k-1)
        call physics_2(B,k)
        B= B+A
```

```
    enddo

Version 2 (3D arrays, serial vectorization):

 real, array(imax, jmax, kmax) :: C, D
 cmf$ layout C(:news, :news, :serial), D(:news, :news, :serial)
 call physics_1_new(C)
 call physics_2_new(D)
 D(:,:,2:kmax)= D(:,:,2:kmax)+ C(:, :, 1:kmax-1)
```

Version 2 is able to take full advantage of the serial optimizations in the compiler (reflected in the use of array syntax for the third (serial) coordinate), but the resulting performance enhancement comes at the cost of factor of 2 x **kmax** in storage. Constructs of this form are very common in SKYHI, and if the full range of serial optimizations is implemented, the result is an explosion in memory usage. This limits the resolution of the models which can be run, affecting both the quality of the numerical experiments which can be performed, as well as the performance, since, on the CM-5, the larger the subgrid, the better the performance of the vector units.

In the early years of parallel programming, much of the discussion relating to different machine architectures tended to center on questions of syntax and the standardization of message-passing protocols. While these issues have not entirely been laid to rest, resolving dilemmas such as the one presented in the previous paragraph will require significant effort in another arena, managing the complex layers of memory, and modes of access to these layers, in the new generation of machines. Concepts such as worksharing, array aliasing, and detailed layouts represent serious attempts by compiler designers to grapple with these issues. For example, a recent extension of data parallel programming, the *global/local* model, allows programmers to take advantage of the data parallel programing style when it suits them, with the additional freedom to drop down to a local (MIMD) level in the middle of their global program. This global/local cycle can be repeated as many times as desired. While it is possible to think of this programming style in terms of an interplay between single and multiple threads of control, it can also be viewed as an extension of the programmer's freedom to manage memory at a fine-grained level. Research is presently underway to investigate the application of new approaches such as these to the memory explosion/optimization tradeoff in SKYHI.

5. Conclusion

It is hoped that the examples and discussion presented in this paper have served to illustrate some of the interesting and complex questions associated with the design of large-scale atmospheric and ocean models for parallel architectures such as the CM-5. At the heart of these questions lies the fundamental issue of communication, whether at the message-passing/run-time level, or the more subtle level implicit in the access to the memory hierarchy of a given machine. Future work will involve trying to find a unified way to manage memory effectively across a range of computer architectures. This will require negotiating the idiosyncracies of hardware and compiler design at a much deeper level than was required for the development of the stencil interface described here. While computational design issues such as these are by no means unique to the field of global climate modeling, the numerical, scientific, and practical challenges presented by the development of global climate models has proved a particularly exciting and productive crucible for our exploration of new conceptual frameworks in parallel program design.

Acknowledgements

I would like to thank Bob Malone and Stephen Boyd for careful readings of this manuscript, and for helpful comments. I would also like to thank the members of the GFDL/LANL collaboration for providing a stimulating and enjoyable atmosphere in which to pursue this work.

References

[1] *Building an Advanced Climate Model,* Program Plan for the CHAMMP Climate Modeling Program, U. S. Department of Energy, Washington, DC 20585 (October, 1990).

[2] *Connection Machine CM-5 Technical Summary*, Thinking Machines Corporation, Cambridge, MA (November 1992).

[3] J. L. Holloway, Jr., and S. Manabe, *Simulation of climate by a global general circulation model: I. Hydrologic cycle and heat balance,* Mon. Wea. Rev. **99**, 335 (1971).

[4] Y. Kurihara and J. L. Holloway, Jr., *Numerical Integration of a nine-level global primitive equations model formulated by the box method,* Mon. Wea. Rev. **95**, 509 (1967).

[5] S. B. Fels, J. D. Mahlman, M. D. Schwarzkopf, and R. W. Sinclair, *Stratospheric sensitivity to perturbations in ozone and carbon dioxide: radiative and dynamical response,* J. Atmos. Sci. **37**, 2265 (1980).

[6] *SKYHI,* in: *Redesign of Research Model Code Sol 52-SAAA-3-00052*, U. S. Department of Commerce/NOAA, Washington D.C. (January, 1993).

[7] C. H. Goldberg and C. L. Kerr, *Development of the Geophysical Fluid Dynamics Laboratory's Climate Models for Scalable High Performance Computer Systems,* Proposal to the Compter Hardware, Advanced Mathematics, and Model Physics Project (January, 1993).

[8] R. D. Smith, J. K. Dukowicz, and R. C. Malone, *Parallel ocean general circulation modeling*, Physica D **60**, 38 (1992).

[9] A. Semtner Jr., *Finite-difference forumulation of a world ocean model,* in: Advanced Physical Oceanographic Numerical Modeling, ed. J. J. O'Brien, (Riedel, Dordrecht, 1986) p. 187.

[10] R. M. Chervin and A. J. Semtner Jr., *An ocean modeling system for supercomputer architectures of the 1990s,* in: Proc. of the NATO Advanced Research Workshop on Climate-Ocean Interaction, ed. M. Schlesinger (Kluwer, Dordrecht, 1988) p. 87.

[11] K. Bryan, *A numerical method for the study of the circulation of the world ocean,* J. Comput. Phys. **4**, 347 (1969).

[12] J. K. Dukowicz, R. D. Smith, and R. C. Malone, *A reformulation and implementation of the Bryan-Cox-Semtner ocean model on the Connection Machine,* J. Atmos. Ocean. Tech. **10**, 195 (1993).

[13] J. K. Dukowicz and R. D. Smith, *Implicit free-surface method for the Bryan-Cox-Semtner ocean model,* J. Geophys. Res., accepted.

[14] M. Metcalf and J. Reid, *Fortran 90 Explained,* (Oxford University Press, New York, 1992).

[15] *High Performance Fortran Language Specification, Version 1.0,* Scientific Programming **2**, no. 1 (June, 1993).

[16] R. C. Malone, personal communication (July 1991).

[17] J. Schwarzmeier and C. Kerr, *Issues associated with domain decomposition for climate models,* unpublished (April 1993).

[18] "Worksharing" refers to an F77-based programming style in which domain decomposition is effected through array scoping attributes ("private", "shared").

[19] The subgrid ratio of a data parallel array executing on a partition of the CM-5 is defined as the product of the extents of all NEWS-type axes (axes which are not in-processor, but which are spread across the machine), divided by the number of vector units in the partition (recall that there are four vector units per PN).

[20] S. R. Atlas, *Performance analysis of the LANL OCEAN code*, unpublished (January 1993).

[21] B. M. Boghosian, *Computational Physics on the Connection Machine,* Computers in Physics **4**, 14 (1990).

EFFICIENCY OF A NUMBER OF PROCESSORS IN INTEGRATING A GLOBAL WEATHER MODEL

S.K.Dash, M.Periasamy*, S.Selvakumar* & B.Jha

Centre for Atmospheric Sciences
Indian Institute of Technology,
New Delhi-110 016, India

* Centre for Development of Telematics, Banglaore, India.

ABSTRACT

The speedup and efficiency of a maximum number of 48 processors are discussed in this paper in the context of integrating a global weather model at different resolutions. It is found that at the present level of parallelisation of the code, the maximum speedup and efficiency achieved for 48 processors are 41 and 85% respectively. Further parallelisation of the model is being attempted to decrease the elapse time for its long term integration on the parallel computer with a view to conduct sensitivity studies concerning climate change.

1. INTRODUCTION

It is a well known fact that numerical weather prediction and climate studies involve very large number of floating point operations and hence supercomputers have been used for the purpose. Global weather forecasting at the leading centres of the world is done on vector supercomputers with a small number of concurrent processors sharing a global memory. For an example, the T213L31 spectral model (about 60kms of horizontal resolution and 31 vertical levels) of European Centre for Medium Range Weather Forecasts (ECMWF) takes about 2 elapsed hours (Dent, 1993) on Cray C90 (using 12 processors) for 10 days forecast. Recent developments in computer technology indicate that in the near future more massive parallelism (Hoffmann & Kauranne, 1993) will be used in weather forecasting and climate studies.

Today, the difficulty in using parallel computers is that the user has to parallelise the source code depending on the application tools, libraries and environment available with the machine. The magnitude of this problem is more in case of weather models which are normally very big in size consisting of thousands of Fortran statements and a couple of hundreds of subroutines. Hence the parallelisation of a weather forecasting model or a climate model and its implementation on a parallel computer is an important task. Before conducting sensitivity studies concerning weather forecast or climate change with the help of a numerical model on a parallel computer, it is worthwhile to exploit as much parallelism as possible so as to achieve the maximum efficiency and to decrease the elapse time for integration.

A global weather model was modified (Dash et al. 1993) for its implementation on 16 processors of C-DOT High Performance Parallel Processing System (CHIPPS) developed by Centre for Development of Telematics (C-DOT) in India and on Meiko T800. It was found that the performance of the model on 16 processors was about 8 times slower than on Cray-XMP with one processor in the case of 10-day integration. This parallel model had the horizontal resolution of T21 (Triangular truncation at 21 waves) and there were 5 levels in the vertical. The horizontal resolution has been increased upto T63 with 9 levels in the vertical. The parallel T63 model has been implemented on 48 processors of CHIPPS. The parallel model at horizontal resolution T42 has also been implemented on all the 32 processors of Meiko. Following Foster et al. (1992) the speedup and efficiency achieved in case of 16, 32 and 48 processors are computed in the present paper. The time taken for all overheads (data transfer, sequential computations and idling) has been reduced considerably with the development

F.-X. Le Dimet (ed.), High Performance Computing in the Geosciences, 257–261.

of an algorithm ('peadd') for CHIPPS. Thus the speedup and efficiency of 48 processors in case of T63 model have been enhanced considerably.

2. C-DOT High Performance Parallel Processing System

The CHIPPS has been designed and fabricated by C-DOT (Periasamy, 1992). It is based on the Single Algorithm Multiple Data (SAMD) architecture. It has the advantages of both Single Instruction Multiple Data and Multiple Instruction Multiple Data architectures. It also provides uniform start-up and simultaneous data distribution across all configurations and it is realised using off-the-shelf hardware and software technology and tools. It is designed to support large, medium and small applications. The ranges include a) large 192 node machine, b) 64 node machine and c) compact 16 node machine to suit various applications depending on the size of the problem.

The Parallel Elements (PEs) are presently based on IMS T800, 32-bit transputers with a peak performance of 2.25 M flops at 25 MHz. Each PE has a local memory of 4 M bytes. A front-end processor (MC68010) has been provided for communication and control. While one high-speed link at 8 Mbps is used to receive program/command broadcast from the MC (Main controller), another is used for data transfer between PE and global memory (upto 6 MW) called Multi-Dimensional access Memory (MDM) through the Inter Connection Network which is a fully non-blocking cross-bar switch.

Application program such as the spectral General Circulation Model (GCM) written for sequential machines can be conveniently parallelised on CHIPPS by identifying the tasks which are independent of other computations. The application program is functionally divided as tasks viz., input/output, data management and computation. The computations are performed at PEs and the control of execution and input/output are done at MC. The user interface is provided through system library to perform tasks like data transfer, switch set-up, execution control etc. The system library, a collection of high level language callable routines, co-ordinates with the kernels at the PEs and MDM and thus helps the user to parallelise the application program in the UNIX environment at the MC.

3. Parallel Model at Resolutions T42 and T63

The details of parallelisation of T21 global model are given in Dash et al. (1993). The sequential code of the model (Dash and Chakrapani, 1989) originated from T21 ECMWF model (Burridge and Haseler, 1977). The parallel codes of the model for resolutions T42 and T63 have been developed from that of T21 model after many modifications. The important changes involve the parameter list, dimension of the input grid point array, work buffers, Legendre array and the spectral array. The basic approach to parallelism is the same as in case of T21 resolution except one important development at the algorithm level so as to reduce the time taken for sequential computations. This is discussed below.

According to the formulation of the global spectral model, the initial data consisting of the meteorological variables are defined at a number of Gaussian grid points (Table 1) depending on the resolution of the model. In order to solve the coupled nonlinear hydrodynamic equations representing the dynamical and physical processes of the atmosphere, the dependent variables are represented by truncated series of spherical harmonics following the triangular truncation scheme in the horizontal plane. Each time step of integration has essentially three stages: the grid point domain, spectral domain and finally the grid point domain. Fast Fourier Transform packages and Legendre Transforms are used for passing from grid point domain to spectral domain and inverse transforms are used for the reverse process. At the current level of parallelism in the model, the grid point computations on a pair of latitudes (one from the Northern hemisphere and the corresponding one from the Southern hemisphere) are done on one PE and thus the number of PEs used for parallel computations all over the globe is equal to half the number of Gaussian latitudes in the model. In the parallel model at resolution T21, the contribution to the spectral co-efficients from each pair of latitudes were computed in parallel on all the PEs and those were sent to the MC (Dash et al. 1993) to be added there in order to get the spectral co-efficients at any timestep of integration. During these additions in the MC, all the PEs were idle. It will be clear in discussions in the next section that this idling time of the PEs contributed immensely to the small speedup and low efficiency of the system.

In the parallel model at resolution T63, a new algorithm has been developed so that it is not necessary to transfer the contributions to spectral co-efficients from all the PEs to the MC for sequential addition. These contributions are exchanged suitably amongst the PEs so that the addition goes on each PE for a particular group of co-efficients depending on the total number of PEs used for model integration. This new algorithm increases the efficiency of the system considerably in integrating the T63 weather model.

4. Speedup and Efficiency

The speedup (Amdahl, 1967) is a measure of time gained in going from one processor to a number of processors for the same problem size. Following Foster et al. (1992), the speedup and efficiency are computed by the following formulae.

$$\text{Speedup,}\quad S = \frac{pt}{t + to} \quad \text{and Efficiency,}\quad E = \frac{S}{p}$$

Here p is the number of processors, t is the time of actual computation on one processor and to is the time for all overheads such as communication of data, sequential calculations and idling of a processor. In order to compute the speedup and efficiency of 16,32 and 48 processors, the following integrations of the model at different resolutions were made. The model at resolutions T21 (15 vertical levels) and T42 (9 vertical levels) was integrated for 30 days using 16 and 32 processors of Meiko T800. The model at resolution T21 with 15 levels in the vertical was also integrated for 30 days using 16 processors of CHIPPS. The model at resolution T63 with 9 levels in the vertical was integrated for one day using 48 processors of CHIPPS. It may be noted that all these integrations were made without the use of new algorithm 'peadd'. In order to demonstrate the advantage of 'peadd' developed for CHIPPS, the T21 model (with 5 vertical levels) was integrated for 10 days using 16 processors and T63 was integrated for 1 day using 48 processors.

In order to avoid computational instability, the model at resolutions T21, T42 and T63 was integrated with time step values 40, 30 and 24 minutes respectively in the semi-implicit scheme. The timings for data transfer, sequential computations and parallel computations were noted separately in order to compute the speedup and efficiency. Speedup and efficiency of 16 and 32 processors for the integration of model for 30 days are shown in Table 2. The results show that the speedup and efficiency of 16 processors for a horizontal resolution of T21 are about 9 and 60% respectively. When 32 processors are used for resolution T42, the efficiency reduces by about half, since the speedup does not change.

Table 3 shows the comparative statement of elapsed timings for parallel computations and other overheads in case of integration of T21 model using 16 PEs and T63 using 48 processors of CHIPPS with and without algorithm 'peadd'. The reduction of time for overheads with the use of 'peadd' is very clear. In case of T63, the reduction of overhead time is five fold. The time for parallel computations has increased considerably due to the use of 'peadd' in case of T63. There is marginal increase of time for parallel components in case of T21. Thus 'peadd' shows very good performance in case of larger problem size. This is also reflected in the speedup and efficiency. In case of T63, the efficiency shows an enhancement from 24% to 85% (Table 3).

Horizontal resolution	Approx. horizontal resolution(km)	No. of latitudes	No. of grids along each latitude	No. of vertical levels	Central memory in K words
T21	600	32	64	5	131
T42	300	64	128	9	288
T42	300	64	128	15	415
T63	200	96	192	5	298
T63	200	96	192	9	451
T63	200	96	192	15	680

Table 1 : Specifications for Different Resolutions of Spectral GCM

Machine	Horizontal resolution of the model	No. of processors	No. of levels levels in the vertical	Speedup	Efficiency in %
Meiko	T21	16	15	9.3	58
Meiko	T42	32	9	9.0	28
CHIPPS	T21	16	5	9.7	61

Table 2 : Efficiency of 16 and 32 Processors for Model Integration for 30 Days.

	T21 for 10 days (16 PEs)		T63 for 1 day (48 PEs)	
	without 'peadd'	with 'peadd'	without 'peadd'	with 'peadd'
Time for Parallel computations (min.)	16.52	16.77	13.23	44.46
Time for overheads (minutes)	11.40	4.80	41.64	8.05
Speedup	9.47	12.44	11.57	40.64
Efficiency (%)	59.17	77.75	24.11	84.67

Table 3 : Efficiency of Chipps using New Algorithm 'peadd'

5. Conclusions

Speedup and efficiency of 16, 32 and 48 processors have been discussed in this paper by integrating spectral GCM with horizontal resolutions T21, T42 and T63 respectively. Maximum number of 32 processors of Meiko T800 and 48 processors (T800) of CHIPPS (built in India) have been used for the purpose of integrating the model upto 30 days. A new algorithm has been developed for CHIPPS so as to make the final addition on all the processors in parallel to get the spectral coefficients of meteorological parameters. Earlier this addition was done in t ıe Main Controller sequentially, thus contributing to low efficiency and small value of speedup. In case of model integration at horizontal resolution T63 for one day, the time for all overheads (data transfer, sequential computations and idling of a processor) has decreased five folds due to the use of this new algorithm. Maximum value of efficiency achieved with 48 processors is 85% and the corresponding speedup is 41. Further optimisation of the parallel code of the model is being attempted so as to enhance the level of parallelism and to decrease the elapse time of integration of global weather model on the parallel computer. This will help in integrating the model on the time scale of a season or more so that various sensitivity studies concerning variability of monsoon and climate change will be carried out using the cost-effective parallel computer.

ACKNOWLEDGMENTS

The CHIPPS project was funded by the Department of Science and Technology, Government of India and Tata Institute of Fundamental Research, Bombay, India. The authors wish to thank Dr.V.K.Misra, Joint Director, DST, Dr.B.D.Pradhan, Executive Director, C-DOT and Prof.M.V.Pitke, TIFR for their support.

REFERENCES

Amdahl G., 1967: The validity of the single processor approach to achieving large scale computing capabilities. AFIPS Conference Proceedings, SJCC, 3, 483-485.

Burridge, D.M. and J.Haseler, 1977: A model for medium range weather forecasting: Adiabatic formulation. Technical Report No.4, ECMWF, Reading, UK, pp.46.

Dash, S.K. and B.Chakrapani, 1989: Simulation of a winter circulation over India using a global spectral model. Proc. Indian Acad. Sci., (Earth and Planet Sci.), 98, 189-205.

Dash, S.K., B.Jha, S.Selvakumar, M.Periasamy, M.V.Pitke and V.K.Misra, 1993: Implementation of a T21 global spectral model on 16 nodes of parallel computers. Parallel Supercomputing in Atmospheric Science, Editors: G-R.Hofftmann and T.Kauranne, World Scientific, 60-72.

Dent, D., 1993: Parallelisation of the IFS model. Parallel Supercomputing in Atmospheric Science. Editors: G-R. Hoffmann and T.Kauranne, World Scientific, 73-88.

Foster, I., W.Gropp and R.Stevens, 1992: The parallel scalability of the spectral transform method. Mon. Weather Rev., 120, 835-850.

Hoffmann, G-R, and T.Kauranne, 1993: Parallel Supercomputing in Atmospheric Science. World Scientific, pp 532.

Periasamy, M., 1992: C-DOT's data parallel computer: An alternative for the 1990s. Horizons, April June 1992, 42-44.

GPSR Compliance

The European Union's (EU) General Product Safety Regulation (GPSR) is a set of rules that requires consumer products to be safe and our obligations to ensure this.

If you have any concerns about our products, you can contact us on ProductSafety@springernature.com

In case Publisher is established outside the EU, the EU authorized representative is:

Springer Nature Customer Service Center GmbH
Europaplatz 3
69115 Heidelberg, Germany

Batch number: 10584568

Printed by Printforce, the Netherlands